# BARRON'S

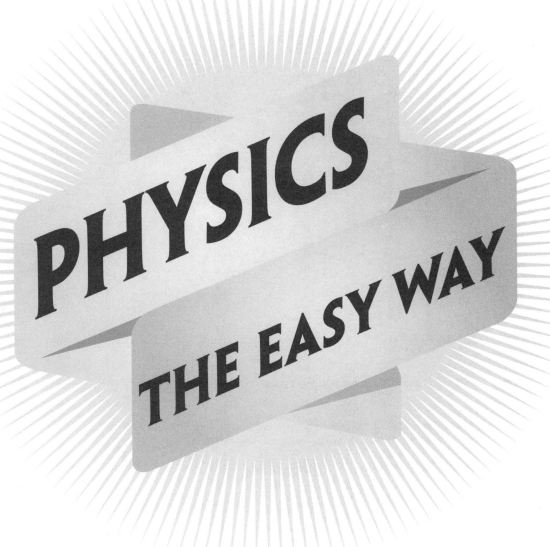

# PHYSICS
## THE EASY WAY

# LEARN PHYSICS THE EASY WAY!

**KENNETH RIDEOUT, M.S.**
*Physics Teacher*
*Science Department Head*
Wayland High School
Wayland, MA

## Acknowledgments

I would like to thank my editors at Barron's for this opportunity to pour my 20-years' experience of teaching physics into this book and thus being able to share my joy and fascination with physics with even more students! I would also like to thank my wife, Irene, and two children, Isabelle and Sebastien, who indulge my enthusiasm for physics with patience and bemusement.

Published by Kaplan, Inc., d/b/a Barron's Educational Series
750 Third Avenue
New York, NY 10017
**www.barronseduc.com**

ISBN: 978-1-4380-1263-6

10 9 8 7 6 5 4 3 2 1

Kaplan, Inc., d/b/a Barron's Educational Series print books are available at special quantity discounts
to use for sales promotions, employee premiums, or educational purposes. For more information or to
purchase books, please call the Simon & Schuster special sales department at 866-506-1949.

# CONTENTS

# PREFACE

## WHAT IS PHYSICS?

Almost all children go through a phase of asking "Why?" repeatedly to the adults around them until the adults are exhausted or annoyed. Physicists are these children who never lost this questioning as they grew up. You can start with ANY subject and ask "Why?" repeatedly. Eventually, when you reach the limits of human understanding, you will inevitably find yourself in the domain of physics. Physics is the oldest discipline of science—the most fundamental in the true sense of the word in that it is looking for those fewest, simplest principles upon which our universe appears to operate. When we ask "Why?" the surprising answer is that we have, through systematic testing over the past 350 years, come up with many answers. Because the question is so vague and phenomena are so varied, the study of physics may appear overwhelming to the new student. The diligent student, however, will be rewarded with an enriched understanding of just about everything in life. The underlying symmetries and rules that result in such a rich personal experience are revealed to the physics students, much like the frame and foundation of a building hidden under its layers of decoration and paint reveal the building's secrets as to how it remains standing.

## WHY STUDY PHYSICS?

Our modern lifestyles are based heavily on modern technology, and all the modern technology we currently enjoy is based on our deep understanding of physics. To understand truly not only the universe we live in but our civilization as well, we should learn to appreciate the basic building blocks of it all: physics. The modern workplace is increasingly technological as well, and many of the as-yet uninvented careers will have some basis in understanding the underlying issues. Even if a given student is no longer interested in the question "Why?" or will not be involved with technology in any way in the future, there is another reason to study physics. Physics is the best means by which to teach and learn the general techniques of problem solving. How do you break down a new problem that you have never seen before into smaller, more approachable pieces? How do you hone your intuition and learn to use the facts you do know as tools to pry out a solution from a difficult problem?

## WHY THIS BOOK?

This book is set up so that students can use it as the primary source to teach themselves physics or as a supplement for students already enrolled in a first- or second-year physics class. As a physics teacher, I have learned that it is really the conceptual scaffolding of physics that students find difficult. Although there are complaints about the mathematical nature of the problem solving, almost invariably, the real issue lies in interpreting the problems and setting them up. With this in mind, this book focuses on the conceptual frameworks and underpinning ideas in physics. Problem solving and details about doing physics are also here—in the examples and the chapter review exercises (with solutions) throughout the text.

Ken Rideout
B.S. Physics with Honors, Purdue University
M.S. Physics (A.B.D.), Carnegie Mellon University

# THREE TYPES OF RELATIVITY

( 1 )

## WHAT YOU WILL LEARN

- All motion is relative.

- The measures of space and time themselves are relative.

- Space and time are distorted by the presence of energy and mass.

- What the correspondence principle is.

| LESSONS IN CHAPTER 1 | |
|---|---|
| • Galilean Relativity | • General Relativity |
| • Special Relativity | |

## 1.1 Galilean Relativity

**Galilean relativity**, also known as classical relativity, is the intuitive sense of the relative nature of motion. If you look at a stationary object outside your window while you are on a moving train, the object appears to be going backward. You intuitively understand that the object is actually standing still while you are moving forward. Only from your point of view does the object appear to be going backward. This idea that motion must be described relative to some arbitrary coordinate system is baked into all of math and physics.

Galilean relativity further supposes that our measurement tools for distance and time remain the same as we change reference points. Look at this in another way. We may all carry around our own graph paper on which to make measurements and our own clocks with which to measure time, but Galilean relativity assumes that the size of the increments on the graph paper are all the same (Figure 1.1) and the ticks of the clock are all the same. All that's needed to translate one person's measurements to another frame of reference is to align the origins of each person's graph paper. (See Section 4.1 for some more details about this process.)

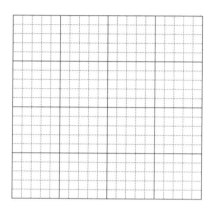

**FIGURE 1.1** Euclidean space-time found in Galilean relativity.

## 1.2 Special Relativity

In 1905, Albert Einstein published a solution to what appeared to be a minor problem in electromagnetism. This new way of thinking about measurements and relative motion would profoundly reshape our thinking about what is really happening in relative motion. The minor problem in electromagnetism was that the mathematics of electric and magnetic fields as embodied in Maxwell's equations were well verified and considered complete, but they give a *specific* speed for electromagnetic radiation. In other words, the speed of light did not appear to be dependent in any way on the observer's own reference frame or coordinate system. This, coupled with recent experiments showing that light did not even need a medium to propagate, brought Einstein to the following two postulates (which seem innocent enough at first).

1.  All inertial observers agree about the laws of physics. Indeed, since Newton's time, all of physics is based on this assumption. The word *inertial* means motion that is not accelerating.

2.  All inertial observers measure the same speed of light (commonly represented by the letter $c$) with a value just under $3 \times 10^8$ m/s. This postulate puts the speed of light on a pedestal among constants: a universal constant that all observers measure to be the same value.

    These two postulates are the core of Einstein's **special relativity**. (This theory is "special" because it covers only the special case of nonaccelerating observers.) Einstein went on to show that the only way all inertial observers could possibly agree on a common value for the speed of light is if they no longer agree about their measuring units. Rather, careful inertial observers that are traveling at different speeds will notice that the other observer is experiencing length contraction and time dilation. That is, another observer moving past you will appear, to you, to have shortened graph paper (in the direction of motion) and a slower clock (that is, time will go by more slowly for the other observer). That observer, in turn, will observe your own lengths and times to be contracted and dilated as well.

Although this seems like a crazy proposition, the actual amount of this contraction and dilation is dependent on the relative speed's fraction of the speed of light (which is a big number: 300,000,000 m/s). For this reason, the effects are not very noticeable in everyday light. Specifically, the factor by which time lengthens and distance contracts is given by the following:

$$\sqrt{\frac{1}{1 - \frac{v^2}{c^2}}}$$

As you can see from the factor, as long as the relative speed ($v$) is much lower than the speed of light ($c$), the effect is minor. As the relative speed between two observers approaches the speed of light, however, this expression approaches infinity! This effect is one of the reasons why we know objects with rest mass can never reach the speed of light relative to another observer and why all observers measure the same speed of light.

Let's return to our graph paper and clock analogies. When we look at another observer's graph paper and clock when that person is in motion relative to us, both the spacing of their graph paper and the ticks of their clock change when compared to our own graph paper and clock (Figure 1.2).

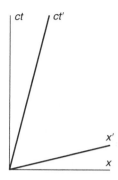

**FIGURE 1.2** In special relativity, an observer in motion (primed axes) will agree with neither distance nor time measurements of a stationary observer.

## 1.3 General Relativity

After resolving how inertial observers must have differing measures of time and space, Einstein turned his attention to accelerating observers. After 10 years of work, he published his findings in what is known as **general relativity**. As with special relativity, Einstein started with a seemingly mild proposition known as the **equivalence principle**:

Inertial mass is the same as gravitational mass.

or

Acceleration and the gravitational field are indistinguishable.

When Einstein followed this assertion to its logical conclusion, he came to a startling conclusion. Not only do the measures of space and time change when traveling at different speeds, but they are warped by the presence of mass and energy. This warping caused by mass is what gravity and inertia are. The little or big "dent" you create in space-time with your mass represents both your inertia (how much you resist changes in motion) and your gravitational influence on objects around you. This major paradigm shift has been tested many times, primarily by looking at the effect that high gravitational fields have on light. In fact, general relativity is one of the most highly tested and verified of all science theories. Later in his life (1952) while reflecting on his amazing discovery, Einstein wrote, "There is no such thing as an empty space. . . . Space-time does not claim existence on its own, but only as a structural quality of the [gravitational] field." In other words, to return to our graph paper and clock analogies one last time, the graph paper and clock exist only because gravity exists (Figure 1.3). In light of this, perhaps we should not be so surprised that the graph paper and clock are so malleable.

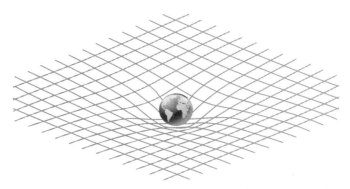

**FIGURE 1.3** Mass distorts the fabric of space-time in general relativity.

Why are these conclusions of Einstein's so unintuitive if experiments have demonstrated their accuracy over our intuition? The answer is simply because we live our lives so locally and at such low speeds compared with the scale of gravitational fields and with the speed of light that we get by just fine by ignoring these effects. However, modern physics would be unable to explain how energy comes from fusion in the sun, how galaxies act as giant gravitational lenses, and how magnetic fields are the same as electric fields if it weren't for both special and general relativity rewriting our assumptions about time and space. It is an important attribute of new, more powerful theories in science that these new theories obey the **correspondence principle**. The correspondence principle simply states that newer theories should give answers that correspond to the older theories under those circumstances. For example, for low relative speeds, special relativity looks and acts like Galilean relativity. For low-mass objects, the distortion of time and space in general relativity is so minor that it can be ignored, and the flat, classical concept of space and time is a fair approximation.

# FORCES

## WHAT YOU WILL LEARN

- Inertia and how it is measured.

- Common forces and how they affect motion.

- The difference between mass and weight.

- Misconceptions in Newton's action-reaction law.

- Drawing and using free-body diagrams.

- Solving net force problems.

| LESSONS IN CHAPTER 2 | |
| --- | --- |
| • Newton's Three Laws of Motion | • Free-Body Diagrams and Solving Force Problems |
| • Types of Forces | |
| • Fundamental Forces | |

## 2.1 Newton's Three Laws of Motion

In 1687, Isaac Newton published what is considered the first physics textbook (commonly referred to as simply *The Principia*) detailing his ideas about the laws of motion and gravity. This kicked off the trend of codifying the rules of nature into mathematical relationships that became known as physics. Although largely supplanted by the modern conservation laws (see Chapter 6) and general relativity (see Chapter 1), Newton's ideas about forces, masses, acceleration, and gravity remain the foundation of what is now called "classical physics." His ideas are an excellent description of how things work in our day-to-day life, which is far from light speed and consists of weak gravitational fields.

Newton's first law of motion is a reversal of the ancient tenet that the natural condition of objects is to come to rest. Instead, based on some earlier work by Galileo, Newton gave us his law of inertia: objects resist changes in their motion. In other words, if objects are already moving, their natural tendency is to continue to move in the same manner. Slide a book across a table and observe it slowing down. In ancient times (Aristotelian thinking), the book is slowing down because the natural state of things is to be at rest. With Newton, the logic is reversed: the book is slowing down because an external force is actively pushing against it. Newton reasoned that if there was no friction between the table and the book, the book would slide in a straight line with no loss of speed. Some objects require more force to *change* their motion; these objects have more **inertia**. Inertia is measured by mass, which has units of kilograms (kg).

Newton's second law of motion tells us how changes in motion happen. If an object slows down, speeds up, or changes direction, an external force must have been applied. A **force** is simply defined as a push or pull. In honor of his work, forces are measured in units of newtons (N) in the modern metric system (SI units—see Appendix B). If several forces are acting, the total (net) force produces this change in motion. A change in speed and/or direction is known as acceleration (units of m/s$^2$). Since an object with greater inertia (more mass) resists changes in motion more than those with less inertia, we are led to the following famous relationship:

$$\text{Acceleration} = \frac{\text{Total force applied to an object}}{\text{Mass of an object}}$$

$$\vec{a} = \frac{\vec{F}_{\text{net}}}{m}$$

From this, it follows that the defined unit of newton is equal to a kilogram meter per second squared:

$$N = kg \cdot m/s^2$$

The notational use of overarrows on a variable is a formal way of indicating it is a vector (as opposed to a scalar). When the overarrow is not present, we are referring to only the magnitude of the variable. For more information about vectors, refer to Appendix A.

Newton's third law addresses the origin of forces. Every force is the result of an interaction between two objects. An important principle of nature is symmetry. Here we have its first manifestation in physics: if one object is pushing or pulling on another object, the exact same force must be acting on both objects but in opposite directions. The two interacting objects creating the force both experience the same force as equal and opposite pushes on each other. This law is, unfortunately, frequently referred to as the action-reaction law. It is one of the most misunderstood laws of physics. It is not a case of one force causing the other; it is not a case of every object having two opposing forces on it; it is not a case of one force coming first and then the opposing force arises in opposition. In fact, it is this:

If object $A$ is exerting a force on object $B$, then object $B$ is simultaneously exerting the same force, in the opposite direction, on object $A$.

Table 2.1 summarizes Newton's three laws of motion. Table 2.2 lists the units used in Newton's laws, both in SI and U.S. customary units.

### TABLE 2.1  SUMMARY OF NEWTON'S LAWS OF MOTION

| Law | Summary |
|---|---|
| First (inertia) | If $\vec{F}_{net} = 0$, motion remains constant. |
| Second (force) | $\vec{a} = \dfrac{\vec{F}_{net}}{m}$ |
| Third (action-reaction) | $\vec{F}_{AB} = -\vec{F}_{BA}$ |

### TABLE 2.2  CHART OF UNITS INVOLVED IN NEWTON'S LAWS

| | SI Units (Modern Metric System) | U.S. Customary Units (Archaic Imperial System) |
|---|---|---|
| Force | newton (N) | pound (lb) |
| Mass | kilogram (kg) | slug |
| Acceleration | meter/second/second (m/s$^2$) | foot/second/second (ft/s$^2$) |

## THINGS TO THINK ABOUT

*Mass vs. weight*

One of the reasons students find the study of physics challenging is that the common usage of many words differs from their scientific usage. For instance, mass and weight are used interchangeably by almost everyone. You might hear someone say, "I weigh 70 kg in Europe but 155 lb in the U.S." This switch between the mass (inertia) of an object and the force due to gravity works only because most people are talking about objects that spend their entire lives in the same constant gravitational field (that is, they remain on the surface of Earth). In actuality, only mass is a property of the object. In contrast, the object's weight depends on the current strength of gravity being experienced by that object. For example, when astronauts were on the moon, they weighed less but had the same mass as when they were on Earth.

All of Newton's laws are dependent on the observer (known as the **reference frame** of the description) being an **inertial** observer. This simply means that the observer is not experiencing acceleration. When an observer employs a point of view that is accelerating, fictitious forces and unexplained accelerations arise (for example, the centrifugal and Coriolis effects).

## 2.2 Types of Forces

Objects can exert many types of pushes and pulls on each other. However, physicists are reductionists. Thus over time, we have been putting the many types of forces discovered into fewer and fewer categories. In fact, for the vast majority of situations you observe directly, you can get by with only five types of forces:

1. Weight

2. Normal force

3. Friction

4. Tension

5. Any push or pull

# EVERYDAY FORCES

**Weight** is the force due to gravity that the Earth (or whichever planet you are on) exerts on your mass. This force is always directed straight toward the center of the planet and is equal to your mass multiplied by the strength of gravity at your location. At sea level on Earth, the strength of gravity ($g$) is around 9.8 N/kg.

$$\text{Weight} = F_g = mg$$

## THINGS TO THINK ABOUT

The definition of weight, $F_g = mg$, is a simplified version of Newton's law of universal gravitation. Students should know the true nature of gravity is that every object with mass attracts every other object with mass very weakly. The farther apart ($d$) the two masses are (when measured center to center), the weaker the gravitational attraction force:

$$F_g = \frac{Gm_1 m_2}{d^2}$$

In this formula, $G$ is a universal gravitational constant equal to $6.67 \times 10^{-11}$ when masses are in kilograms and distance is in meters. The simplified version works as long as you are interested in only the gravitational attraction between an object on the surface of Earth and the planet Earth. In that case, $m_2$ becomes the mass of the Earth and $d$ becomes the radius of the Earth ($R_{\text{Earth}}$). Combining equations gives:

$$F_g = m_1 \left( \frac{Gm_{\text{Earth}}}{R_{\text{Earth}}^2} \right) = m_1 g$$

Solving for $g$ results in the following:

$$g = \frac{Gm_{\text{Earth}}}{R_{\text{Earth}}^2}$$

Plugging in the appropriate values for $G$, $m_{\text{Earth}}$, and $R_{\text{Earth}}$ results in the 9.8 N/kg value used for gravity at the surface of Earth.

The **normal force** ($F_N$) on an object is the force exerted by a surface to keep the object from pushing through the surface. The normal force is a response to the amount of force applied to the surface. The direction of the normal force is always perpendicular (hence the name "normal") to the surface itself. Students sometimes have trouble believing in this force or forget to put it into their analysis of a situation. However, ask yourself, why do your feet hurt after standing all day? Why doesn't gravity pull you through the floor?

The force of **friction** ($F_f$) is applied as matter slides by or bounces off the surface of an object as the matter moves or attempts to move. The direction of the frictional force is always opposite the direction of the relative motion of the two surfaces. If the two surfaces are not yet sliding past each other, the direction of the force is opposite the intended motion. Recall from Newton's third law that both surfaces will experience the same (but opposite) frictional force.

 ## THINGS TO THINK ABOUT

### A deeper look at friction

Frictional forces between objects are a function of velocity. If one or both of those objects is a fluid, such as air, the friction is a bit hard to model. However, the friction between two solid surfaces is much easier to model. Look at the data generated in Figure 2.1 as a block of wood is slowly pulled across a surface until it finally begins to move and then is pulled at constant speed.

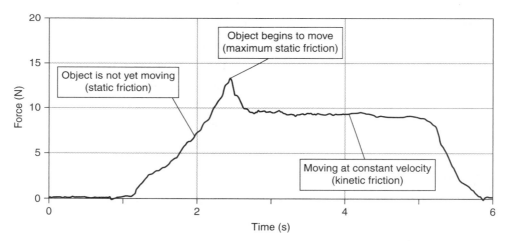

**FIGURE 2.1** Static and kinetic frictional forces

The frictional force arises from millions of atomic repulsions at the surfaces of the two materials. The harder the two surfaces are pressed together, the more the atoms repel each other. This force has already been described—it is the normal force. The other factor in this surface-to-surface friction is the exact nature of the surfaces involved. (What kind of atoms are the surfaces made of? Are they rough or smooth? And so on.) Standard practice is to use a dimensionless coefficient $\mu$ (mu) to represent the nature of these two surfaces.

Note from the graph in Figure 2.1 that before the object begins to move, the frictional forces rise from zero to some maximum value in a linear fashion. This is known as static friction:

$$F_s \leq \mu_s F_N$$

Once the object is in motion, the friction levels off and is a constant value. This is known as kinetic friction:

$$F_k = \mu_k F_N$$

**Tension** is the force that pulls on an object through a string, rope, or other supporting structure attached to the object. Tension is always directed along and through the string, rope, etc., and away from the object at the point of contact. The tension is the same throughout the rope as long as the rope is considered ideal (low mass, not stretching, and not experiencing friction).

Any **push** or **pull** on an object is a force. If you must model a force that does not fit into one of these given categories, label it simply $F_p$.

# OTHER FORCES

In addition to the everyday forces previously listed, other forces play an important role in physics:

- Forces exerted by springs or other elastic object (see Chapter 11), $F_s$

- Electric forces (see Chapter 7), $F_E$

- Magnetic forces (see Chapter 9), $F_B$

- Strong nuclear force and weak nuclear force, which are deep inside of atoms and were discovered in the early 20th century (see Chapter 15)

# 2.3 Fundamental Forces

After looking over the previous descriptions and lists of forces, one might question the reductionist nature of physicists! However, a close inspection of the nature of these forces reveals an underlying feature common to all forces other than gravity: they depend on charges. All macroscopic interactions other than gravity are due to the interaction of charged objects. Take the normal force for instance. What exactly is stopping the two surfaces from going through each other? The outermost layer of matter on each side of each object is made of electrons. As the surfaces are brought close together, the electrons in one surface repel the electrons in the other surface. In the modern view of physics, there are only four fundamental forces, which are neatly divided into two categories:

- Gravity (described by general relativity)

- Electromagnetic force, weak force, and strong force (described by the standard model; see Chapter 15)

Ever dissatisfied, physicists continue to try to unite gravity with the rest of the forces in the standard model in various proposals of some kind of grand unification theory. Thus far, no clear consensus has emerged.

# 2.4 Free-Body Diagrams and Solving Force Problems

A common problem in physics is to relate the acceleration being experienced by an object to the forces being applied. Many students will try to get by with just recognizing certain common situations, but the permutations of these types of problem are endless. In physics, the point is not to get a specific answer to a specific problem. It is more powerful to find a procedure that you can follow in all situations, resulting in a solvable math problem. Since what you are going to learn in this section involves vectors, if you are not familiar with the rules and special vocabulary of vectors, please familiarize yourself with Appendix A now.

Since all forces are vectors, their direction must be considered when determining the net force. Almost all problems in an introductory physics class will be two-dimensional (indeed, even most real-life problems can be reduced to two dimensions). An important tool in solving force problems is the free-body diagram (also known as a force diagram). A free-body diagram strips all superfluous information away from the problem to begin the translation of a description or picture into a math problem. The first step is to represent the object as a simple dot. This dot can be thought of as the center of mass of the object. Next, draw one arrow per force, starting from the dot. Each arrow should be pointed in the proper direction and have its proper label. If possible, attempt to make the lengths of the arrows proportional to the magnitude of the force. Accelerations and velocities are not forces and so should not appear in the force diagram.

**EXAMPLE**    A man is pushing a couch across the floor at a steady pace. He is pushing down on the couch at a 45-degree angle. Draw a free-body diagram for the couch.

**SOLUTION**

First identify all the force being experienced by the couch: the push, Earth's gravity, the normal force of the floor, and the friction between the floor and the couch. Finally, draw an arrow for each force and label each.

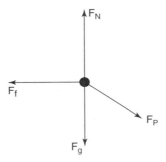

Once you have constructed a free-body diagram for the situation, you can lay down your x- and y-axes with the origin on the center of the free-body diagram. Initially, students always make the x-axis horizontal and the y-axis vertical. However, you can orient these axes in any manner as long as the x- and y-axes are perpendicular to each other. If you know the direction of acceleration, then one of the two axes should be in the direction of the acceleration as this will simplify the math later on. Now it is time to evaluate the net force and relate this to the acceleration. The best way is to write down Newton's second law twice: once for the x-components and once for the y-components. You will have as many terms on the left-hand side as you do forces:

$$F_{x_1} + F_{x_2} + \cdots = \Sigma F_x = ma_x$$
and
$$F_{y_1} + F_{y_2} + \cdots = \Sigma F_y = ma_y$$

At this point, invariably one of the components of acceleration will be zero if you oriented your axes correctly. In fact, usually many of the force components will also be zero. Since there are two equations, the problem must specify all information with the exception of two variables in order to have a solvable problem.

## THINGS TO THINK ABOUT

*Tip*

If you find yourself with too many unknowns, try rereading the problem carefully. Here are the most often overlooked and problematic issues.

- One or both components of acceleration are zero. This occurs when the object is stationary, the object is moving at constant velocity, etc.

- Use the model for sliding friction being proportional to the normal force only if you must. In this case, mu ($\mu$) is given, requested, or implied.

- The problem expects you to ignore friction because the surface is slick, smooth, or slippery or because the object is in free fall.

- Students often put extraneous forces into their free-body diagram. Do not add a new force in the direction of acceleration. Use normal forces only if the object is in direct contact with a surface. Do not double count a single force (e.g., the tension of the line is the pull, not a pull plus a tension). The existing velocity of the object is not a force—that's inertia!

Now, you simply substitute in the given values and solve for the unknowns. One of the most common errors of all is to assign arbitrarily a vertical normal force to be equal to the object's weight, $mg$, by thinking that the normal force's job is to oppose gravity. However, the normal force is found by solving the system of equations generated by the free-body diagram. A diligent student will find that the normal force is equal to $mg$ only when there are no additional vertical forces (beyond gravity and normal) and when there is no vertical acceleration!

A second point of confusion is the sign that should be used for gravity. (Is it +9.8 or −9.8?) Gravity itself is neither positive nor negative. So the student must choose the sign of all components for all forces (including gravity) in the free-body diagram. The first arrow determines the direction of all other arrows. Whether a particular direction (up or down) is positive or negative is a choice made by the student when solving the problem.

Once the student has oriented the $x$- and $y$-axes relative to the free-body diagram, the positive and negative directions are defined. Students should be careful to be consistent once they begin a problem. For example, if you choose to make the downward direction positive in order to have the force due to gravity be a positive number, you are locked into making the upward force component negative.

EXAMPLE Let's return to the previous example. The couch has a mass of 45 kg. If the man is pushing the couch with 175 N of force while it is sliding at a steady pace, determine the normal force experienced by the couch as well as the coefficient of kinetic friction between the floor and the couch.

**SOLUTION**

Assign the $x$-axis to be horizontal to the right and the $y$-axis to be vertically upward. Use Newton's second law for this situation:

$$F_{p,x} + F_{N,x} + F_{f,x} + F_{g,x} = ma_x$$

and

$$F_{p,y} + F_{N,y} + F_{f,y} + F_{g,y} = ma_y$$

Since all motion is along the floor, we can assume there is no $y$-motion much less a $y$-component of acceleration, so $a_y = 0$. Also, since the problem specifies the horizontal motion is at a steady pace, we can assume constant velocity and therefore $a_x = 0$ as well. Putting in the known values, our two equations reduce to:

$$(175)(\cos 45°) + 0 - F_f + 0 = 0$$

and

$$(-175)(\sin 45°) + F_N + 0 - (45)(9.8) = 0$$

Solve for our two unknowns:

$$F_f = 124 \text{ N}$$
$$F_N = 565 \text{ N}$$

Since the problem requested the coefficient of friction as well, we use the model of sliding friction:

$$\mu = \frac{F_f}{F_N} = 0.22$$

# SOLVING FORCE PROBLEMS

1. Draw a free-body diagram showing all the forces acting on the object.

2. Choose a coordinate system. Align one axis with the direction of acceleration.

3. Break any diagonal forces into their horizontal and vertical components.

4. Add all the vertical vector components to determine the net vertical force component.

5. Add all the horizontal vector components to determine the net horizontal force component.

6. Use $\vec{F}_{net} = m\vec{a}$ to relate the components of acceleration to the net force.

# Chapter Review Exercises

1.  If Newton's third law is true, how does anyone win in a tug-of-war competition?

2.  What is the common issue with these expressed impressions?

    In a car turning to the left: "I felt as if I was being pushed outward, to the right."

    In a bus where the brakes have been applied suddenly: "I felt as if I was being thrown forward."

    In a plane during takeoff: "I felt as if I was being pushed back into my seat."

3.  James says, "Newton's third law cannot always be true; for example, what about gravity? If planet Earth is pulling me down with 500 newtons of force, I would notice the Earth is also being pulled up with 500 newtons of force, and I don't see that!" How could you explain to James that the third law is true even in this case?

4.  If an adult weighs 800 N on Earth's surface, what is his weight and mass on the moon (where gravity is six times weaker)?

5.  What is the normal force in the following situations involving a 2 kg book?

    a.  The book sitting on the floor

    b.  The book sitting on the floor while your teacher leans on it with 12 N of force

    c.  The book sitting on the floor while your teacher attempts to lift it with 8 N of force

    d.  The book sitting on the floor of an elevator while the elevator accelerates upward at 2 m/s$^2$

    e.  The book sitting on the floor of an elevator while the elevator moves downward at a constant 2 m/s

6.  A mom is pulling her two children on a sled (total mass = 85 kg) with a cord that makes a 60-degree angle with the horizontal. The sled is moving across a horizontal surface that has a coefficient of kinetic friction of 0.12. Find the normal force and the acceleration of the sled.

# LINEAR MOTION

## WHAT YOU WILL LEARN

- How speeding up, slowing down, and turning are all about the relationship between velocity and acceleration.

- How to calculate average velocity and average speed.

- Slopes and areas in kinematics graphs.

- Problem solving when undergoing constant acceleration.

| LESSONS IN CHAPTER 3 | |
| --- | --- |
| • Acceleration and Velocity Vectors | • Equations of Motion for Constant Acceleration |
| • Distance and Displacement | |
| • Kinematics Graphs | |

## 3.1 Acceleration and Velocity Vectors

In Chapter 2, we defined acceleration as the change in the motion of an object. A more formal definition of acceleration can be obtained using the following relationship:

$$\vec{a} = \frac{\Delta \vec{v}}{\Delta t}$$

Acceleration is equal to the change in velocity divided by the time to change. Note that velocity is a vector and therefore is more than just speed. Velocity is speed and direction, so a change in either is an acceleration. Acceleration, in turn, is also a vector and has the same direction as the change in velocity:

$$\Delta\vec{v} = \vec{v}_f - \vec{v}_i$$

## THINGS TO THINK ABOUT

*Tip*

Students sometimes forget that a change in direction is a change in velocity, and therefore, a net force must be causing a nonzero acceleration. For example, a car takes a turn at a constant speed—is it accelerating? Yes, the car is accelerating since the velocity vector is in a different direction!

 A car ad claims that a car can go from zero to 55 mph in 5 seconds. What is the acceleration of the car in m/s²?

**SOLUTION**

$v_i = 0$

$v_f = 55$ miles/hour

$\Delta t = 5$ seconds

First do dimensional analysis (see Appendix B) to change the units:

$$v_f = \frac{55 \text{ miles}}{\text{hour}} \times \frac{1{,}609 \text{ meters}}{\text{mile}} \times \frac{1 \text{ hour}}{3{,}600 \text{ seconds}} = 24.6 \text{ m/s}$$

Then calculate the acceleration:

$$a = \frac{\Delta v}{\Delta t} = \frac{(v_f - v_i)}{5 \text{ s}} = \frac{(24.6 - 0) \text{ m/s}}{5 \text{ s}} = 4.9 \text{ m/s/s} = 4.9 \text{ m/s}^2$$

When put in terms of *g* (the acceleration we all are familiar with from Earth's gravity):

$$4.9 \text{ m/s}^2 = (0.5)(9.8 \text{ N/kg}) = 0.5g$$

Note that the words and phrases "speeding up," "slowing down," and "turning" all indicate an acceleration but do not indicate a sign (+/−) for the acceleration. What can be inferred are the relative directions of the acceleration and velocity vectors. If the acceleration and velocity vectors are in the same direction, as shown in Figure 3.1, the object is speeding up. If those vectors are in opposite directions, as in Figure 3.2, the object is slowing down. If the acceleration and velocity vectors are at right angles to each other, as shown in Figure 3.3, the object is turning.

**FIGURE 3.1** Case 1: Acceleration (light blue) and velocity (dark blue) vectors are in the same direction. The object is speeding up.

**FIGURE 3.2** Case 2: Acceleration (light blue) and velocity (dark blue) vectors are in opposite directions. The object is slowing down.

**FIGURE 3.3** Case 3: Acceleration (light blue) and velocity (dark blue) vectors are at right angles. The object is turning.

Of course, the acceleration vector can be at some other angle relative to the velocity vector. In that case, the object will be both turning and speeding up (or turning and slowing down). In this chapter, we will be dealing with linear motion and therefore only the first two cases, shown in Figures 3.1 and 3.2, are used in this chapter. Refer to Chapter 4 for problems that involve changes in direction.

# 3.2 Distance and Displacement

In addition to the distinction of velocity and speed, displacement and distance are also words commonly used interchangeably by nonphysicists but are different in important ways:

- Displacement is a vector connecting your starting point to your final position (points from the start toward the finish).

- Distance is a scalar and includes all the ground you've covered, regardless of direction.

EXAMPLE   You take your dog out for a walk around the block. Each side of the square is 200 meters long. What is the total distance traveled? What is your displacement?

**SOLUTION**

The distance is 800 meters. The displacement is zero because your final position is the same as your initial position.

Now that distance and displacement have been defined, the definitions of average speed and average velocity can now be presented:

- Average speed is $\dfrac{\text{total distance}}{\text{total time}}$.

- Average velocity is $\dfrac{\text{net displacement}}{\text{total time}}$.

In addition to describing the average speeds and velocities for some interval of time, we can also talk about the speed and velocity at an instant in time. These are called, respectively, instantaneous speed and instantaneous velocity. Instantaneous speed is the absolute value (or magnitude) of instantaneous velocity.

# 3.3 Kinematics Graphs

Velocities have a very important relationship to position vs. time graphs, as shown in Figure 3.4.

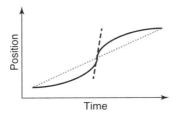

**FIGURE 3.4** Positive vs. time graph.

If the dark blue line in Figure 3.4 represents the actual path followed over time, it is not easy to see what the instantaneous speed is at any particular time. However, the displacement is easily found by simply connecting the starting location to the final location (light blue line). The slope of this light blue line connecting start and finish is $\frac{\text{displacement}}{\text{time}}$. So slopes on position vs. time graphs are actually velocities! An instantaneous velocity can be found by drawing in a tangent line to the curve at a particular point. For example, the slope of the dashed line would be the instantaneous velocity at the halfway point of this trip.

This trick of looking at the slope of graphed data is a common one in physics. (Just look at the units of the *y*-axis over the units of the *x*-axis to see if they mean anything in particular.) We continue the graphical analysis of kinematics by examining the velocity vs. time graph shown in Figure 3.5.

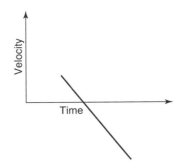

**FIGURE 3.5** Velocity vs. time graph.

Here, the velocity of a ball tossed straight up into the air is shown. Note that the velocity goes through zero at the ball's highest point and then the velocities are negative afterward as the ball is on its way down. This linear graph has a constant negative slope. What is the meaning of this slope? It's the acceleration (change in velocity/time)! In this example, the slope would give the acceleration due to free fall on Earth, which, of course, is a constant value (hence the straightness of the graph) as gravity does not change in any way during the flight of the ball.

There is one more trick for graphs, which is often not taught until a student has taken a course in calculus. The area under a curve may also be physically meaningful. (The units of that area are the units of the two axes multiplied by each other.)

Table 3.1 gives a brief review of the information we can extract from the various graphs of motion.

**TABLE 3.1 INFORMATION EXTRACTED FROM VARIOUS GRAPHS OF MOTION**

| Type of Graph | Information Extracted |
|---|---|
| Displacement vs. Time | Slopes are velocities. |
| Velocity vs. Time | Slopes are accelerations. |
|  | Areas under the curve are changes in displacement, $\Delta d$. |
| Acceleration vs. Time | Areas under the curve are changes in velocity, $\Delta v$. |

EXAMPLE    Describe the velocity and acceleration of the object pictured below.

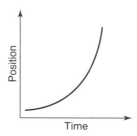

**SOLUTION**

Although the velocity is always positive, it is increasing as time goes on, as can be seen by the increasingly steep positive slope of the curve. Since the velocity is positive and is getting bigger, there is an acceleration, which is also positive.

**EXAMPLE** What is the displacement of the object during these 5 seconds?

What is the average acceleration over this 5-second interval?

What is the instantaneous acceleration at the 2-second mark ($t = 2$)?

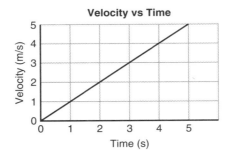

**SOLUTION**

Displacement is the area under the curve, which is a triangle:

$$\frac{1}{2}(5 \text{ s})(5 \text{ m/s}) = 12.5 \text{ m}$$

Average acceleration is the slope of the line:

$$\frac{\text{Rise}}{\text{Run}} = \frac{(5 - 0) \text{ m/s}}{5 \text{ s}} = 1 \text{ m/s}^2$$

Instantaneous acceleration along this graph is the same as the average slope since the graph is a straight line: $1 \text{ m/s}^2$.

# 3.4 Equations of Motion for Constant Acceleration

The graphical principles given above are the student's best tools for the general case of relating position, velocities, and accelerations. However, the special case of constant acceleration (straight lines in velocity vs. time graphs) lends itself to algebraic manipulations. From the two properties of these graphs, we can derive our two general equations of motion:

Area under the graph of velocity vs. time:

$$\Delta d = v_i t + \frac{1}{2}at^2$$

Equation of a line:

$$v_f = v_i + at$$

Table 3.2 shows the five kinematic variables for constant acceleration problems.

**TABLE 3.2  FIVE KINEMATIC VARIABLES**

| | |
|---|---|
| Initial velocity | $v_i$ |
| Final velocity | $v_f$ |
| Displacement | $\Delta d$ |
| Acceleration | $a$ |
| Time | $t$ |

In general, we have problems where we are given some information and asked to solve for the unknowns using our equations of motion. Since we have two equations, the problem is solvable if there are only two unknowns. Since there are five variables in the two equations, we are left with an important rule for kinematics problems involving constant acceleration: at least three numbers must be specified or implied by the problem.

In physics, the hardest job is often translating a given problem into a well-defined math problem. Frequently students will rush to solve the problem by jumping into mathematical manipulations right away. However, time must be spent up front in identifying the given information carefully and being clear about what is being requested.

First decide where your origin will be located. Second, decide which direction you will call positive. Third, identify all the given information (with the correct sign!). If you have not identified three values, you should not proceed until you reread the problem. There are several ways to rearrange these equations in order to eliminate one of the five variables (the most commonly presented one eliminates time: $v_f^2 = v_i^2 + 2a\Delta d$). However, all problems can be solved from just the two basic equations given above.

 **THINGS TO THINK ABOUT**

*Commonly implied values in kinematics problems*
"begins from rest" or "dropped" ($v_i = 0$)
"comes to rest" or "stops" ($v_f = 0$)
"free fall" or "falling freely" ($a = -9.8 \text{ m/s}^2$)
"constant velocity," "cruising," and "steady pace" ($a = 0$)

# Chapter Review Exercises

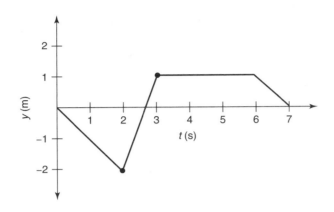

1. Determine the average velocity from the graph above:

   a. During the first 5 seconds

   b. During the first 2 seconds

   c. From the second to the seventh second

2. Determine the instantaneous velocity from the graph above:

   a. At the first second

   b. 2.5 seconds into the trip

   c. At the fourth second

3. What is your average velocity if you drive due north for 20 minutes at 35 mph, then stop for exactly 1 hour, and then drive due south for 10 minutes at 40 mph?

4. Determine your average speed for the trip in Problem 3.

5. You throw a rock straight up in the air at +20 m/s and then catch it when it comes down. Use + to indicate the up direction and use – to indicate the down direction, and do not forget your units.

a. What is the velocity of the rock when you catch it?

b. What is the velocity of the rock at its highest point?

c. How long is the rock in flight?

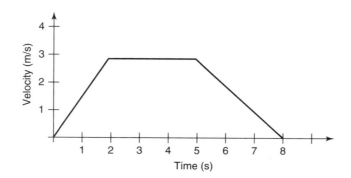

6. Determine the average acceleration from the graph above:

a. During the first 5 seconds

b. During the first 2 seconds

c. From the second to the eighth second

7. Determine the instantaneous acceleration from the graph above:

a. At the first second

b. 2.5 seconds into the trip

c. At the seventh second

8. Determine the net displacement for the trip in the graph above.

9. If a car slows from 25 m/s to 9.0 m/s in 4 seconds, what is its average acceleration?

10. For problem 9 above, how many meters does the car travel in the 4 seconds?

# 4

# 2-D MOTION

## WHAT YOU WILL LEARN

- How to change a moving reference frame for velocities.

- Using Newton's second law when changing direction.

- Projectile motion is free fall with constant horizontal velocity.

- Orbital motion is circular motion with gravity acting as the centripetal force.

- How astronomers determined the mass of the sun and other astronomical objects.

| LESSONS IN CHAPTER 4 | |
|---|---|
| • Relative Velocity | • Projectile Motion |
| • Centripetal Forces and Circular Motion | • Orbital Motion |

## 4.1 Relative Velocity

How fast are you moving right now? If you are seated somewhere while reading this, you might be tempted to say 0 km/hr (or 0 mph if you live in one of three countries that have not yet adopted the metric system). However, what you mean is 0 km/hr relative to the ground beneath your feet. In reality, of course, the ground is moving (probably several hundred km/hr, depending on your latitude) as Earth completes one rotation every 24 hours. Your speed right now is even higher if you take the sun as your reference point (which means you're traveling over 100,000 km/hr). Which number is most correct?

Of course, they are all equally correct—there is no such thing as absolute velocity in our universe: all motion is relative.

If all motion is relative, how do you translate one relative velocity into another? The answer is vector addition.

Let $\vec{v}_{ab}$ mean the "velocity of $a$ relative to $b$." Then relative velocities are translated as follows:

$$\vec{v}_{ab} + \vec{v}_{bc} = \vec{v}_{ac}$$

This equation means that the velocity of object $a$ relative to $b$ plus the velocity of object $b$ relative to $c$ results in the velocity of object $a$ relative to object $c$.

Note the order of the subscripts is important! If you have a velocity description that is backward, you can use the following trick to reverse the order of the subscripts:

$$\vec{v}_{ab} = -\vec{v}_{ba}$$

**EXAMPLE**  To reach a destination on time, a pilot must fly southeast at 250 km/hr, but there is a wind of 75 km/hr blowing 30 degrees north of east. What speed and direction should the pilot head the plane in the air such that after the wind changes the plane's course, the speed and heading will be correct?

**SOLUTION**

First identify the vector quantities cleanly in the format for relative velocities:
Wind speed = $v_{ag}$ (velocity of air with respect to ground)
Final/resultant heading = $v_{pg}$ (velocity of plane with respect to ground)
Initial heading = $v_{pa}$ (velocity of plane with respect to the air)

Next place these terms in their relative velocity relationship:

$$\vec{v}_{pa} + \vec{v}_{ag} = \vec{v}_{pg}$$

Solve for the unknown:

$$\vec{v}_{pa} = \vec{v}_{pg} - \vec{v}_{ag}$$

Break your vector equation into the corresponding component equations:

$$\vec{v}_{pa,x} = \vec{v}_{pg,x} - \vec{v}_{ag,x}$$

$$\vec{v}_{pa,y} = \vec{v}_{pg,y} - \vec{v}_{ag,y}$$

Find the components of the given vectors and solve:

$$\vec{v}_{pa,x} = (250\cos45°) - (75\cos30°) = 112 \text{ km/hr}$$

$$\vec{v}_{pa,y} = (-250\sin45°) - (75\sin30°) = -214 \text{ km/hr}$$

Finally, convert the component results to a single magnitude and angle:

speed = 242 km/hr

62.4 degrees south of east

You can also graphically add the vectors to solve the problem (see Appendix A), as shown below.

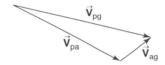

# 4.2 Centripetal Forces and Circular Motion

Recall from Chapter 3 that simply changing direction does indeed produce a new velocity vector (it may have the same length, but it has a new direction). So acceleration is associated with changes in direction. In turn, these particular accelerations require a force that is perpendicular to the velocity. If the perpendicular force moves with the object as it turns, the object will continue to turn and create a curved path. If the curve continues for long enough, a complete circle may be formed. All smooth curves can be approximated as an arc of a larger circle, and thus these types of analyses in physics are known as **circular motion problems**. Circular motion problems do not fit nicely into the usual horizontal and vertical directions. Instead, it is more useful to indicate directions that are radial (outward or inward) and tangential:

**Centripetal** means *center seeking.*

**Centrifugal** means *center fleeing.*

Note that the radial direction and the tangential direction (indicated respectively by the acceleration and velocity vectors in Figure 4.1) are at right angles to each other like our usual $x$- and $y$-axes.

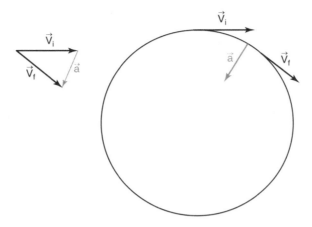

**FIGURE 4.1** Acceleration and velocity vectors.

The velocity vector is always tangent to the curved path, and the acceleration vector is always inward. Hence the force (and acceleration) required is centripetal rather than centrifugal. Note that a net centrifugal force would actually cause an outward change in direction and the motion would be entirely different. Since the net force required is centripetal, the sign convention when determining the net force in circular motion problems is that inward forces are positive and outward forces are negative (recall the tip in Chapter 2 to always align your positive axis with the direction of acceleration). Tangential forces (or components of forces) will speed up or slow down the motion just the same as in one-dimensional motion. The centripetal acceleration ($a_c$) required to stay on the curve is a function of current speed ($v$) and the radius of the curve ($r$):

$$a_c = \frac{v^2}{r}$$

Note that although $v$ is traditionally used in these formulas involving circular motion, once a vector is squared, all directional information is lost. Thus, in this particular case, speed and velocity are interchangeable. Newton's second law for circular motion in the radial direction becomes:

$$F_{net} = F_{inward} - F_{outward} = ma_c = \frac{mv^2}{r}$$

## THINGS TO THINK ABOUT

Centripetal and centrifugal are not new types of forces! They are just adjectives describing the *direction* of the usual forces. Any force can act centripetally (or centrifugally if it is pulling you outward).

If centrifugal forces are not required (indeed are antithetical) to circular or curved motion, why is the term so commonly used? The **centrifugal effect** is an illusion caused by Newton's first law.

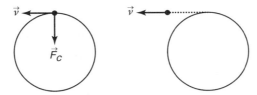

**FIGURE 4.2** If the centripetal force is taken away,
the object flies out along the tangent line

The object undergoing circular motion has inertia along the tangent line since that is the direction of the velocity vector (see Figure 4.2). The fact that some force must push you toward the center, along with Newton's third law, makes you feel as if you had been thrown out of the circle! Take the common case of being a passenger in a car that is taking a sharp right-hand turn: you will feel as if you are being pushed to the left from your accelerating reference frame. However, what is actually happening from the inertial reference frame of the ground is that your body is continuing in a straight line path as the car moves inward on you until the car (via the door or the seatbelt) applies inward forces on you to make you turn with the car. Of course, since the car is applying an inward force on you (centripetal), by Newton's third law you are applying an equal but outward (centrifugal) force on the car. Also adding to the impression of a centrifugal effect is that when objects don't experience sufficient inward forces to keep them on the circular path, they fly off along the tangent line, which puts them outside the circle. Note the path they follow is not radial (which would be centrifugal) but rather tangential (just as Newton's first law demands).

## THINGS TO THINK ABOUT

Is it surprising that passengers feel an outward sensation during a turn when the actual force is inward? Before you answer, think about the sensation they feel during a sudden forward acceleration and the sensation they feel during sudden braking (backward acceleration). In all cases, the impression of the passengers is in the opposite direction of the actual acceleration (the difference between where they expected to be due to your inertia and where they actually are due to the acceleration).

**EXAMPLE** Although it is not possible to create artificial gravity for astronauts on a space station, one trick is to put them into a rotating structure and have them "stand" on the inside surface of the outer wall. The inward normal force could be made to be as strong as the normal force they feel on Earth due to their weight. To create an artificial gravity similar to that on Earth, how many rpm (revolutions per minute) must a cylindrical space station (radius = 85 meters) rotate for astronauts standing on its inner surface to feel like they are on Earth?

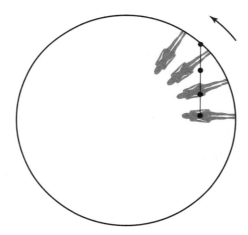

### SOLUTION

The normal force on the space station is the centripetal force:

$$F_N = F_c = ma_c = \frac{mv^2}{r}$$

You want the situation when $F_N = F_g = mg$:

$$mg = \frac{mv^2}{r}$$

Solve for $v$:

$$v = \sqrt{gr} = \sqrt{9.8 \times 85} = 29 \text{ m/s}$$

To determine rpm, use $2\pi r = 1$ revolution and 60 seconds = 1 minute:

$$\frac{29 \text{ meters}}{\text{second}} \times \frac{1 \text{ revolution}}{2\pi r} \times \frac{60 \text{ seconds}}{1 \text{ minute}} = 3.3 \text{ rpm}$$

# 4.3 Projectile Motion

Projectile motion is the 2-D version of free fall (the only force acting on an object is gravity). In addition to experiencing the vertical motion under gravitational acceleration, the object is also experiencing a constant horizontal velocity, as seen in Figure 4.3.

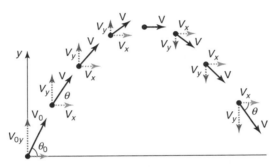

**FIGURE 4.3** Projectile motion.

The key to analyzing projectile motion is to divide the problem into two problems: one horizontal and one vertical. We will use the convention that upward is the positive $y$-direction and the horizontal direction of the launch is the positive $x$-direction.

**Horizontal ($x$)**      **Vertical ($y$)**

$a_x = 0$                 $a_y = -g$

                           $(g = 9.8 \text{ m/s}^2)$

Generally, a launch speed ($v_0$) and angle ($\theta$) are given:

$v_{xi} = v_0\cos\theta$          $v_{yi} = v_0\sin\theta$

(constant)

Our equations of motion give us:

$$v_x = v_{xi} \qquad\qquad v_y = v_{yi} - gt$$

$$\Delta d_x = \Delta x = v_{xi}t \qquad \Delta d_y = \Delta y = v_{yi}t - \frac{1}{2}gt^2$$

To answer any question about the range ($x$), altitude ($y$), or velocity, you must first determine the time in flight to the point in question. Usually, you will determine the time from the vertical equations. Also remember that at the highest point in the object's flight, $v_y = 0$.

If the parabola is symmetrical (same initial and final heights), the time in flight is double the time to the highest point. In addition for a symmetrical parabola, landing speed is the same as launching speed.

If a specific height is known ($\Delta y$ has a value), the resulting quadratic equation for time must be solved. Be careful with your sign for $\Delta y$ (low to high = positive; high to low = negative)! Negative roots for time can be thrown away.

Speed at any point during flight is found by "Pythagorizing" the components of velocity, $v_x$ and $v_y$, by using $\left( \sqrt{v_x{}^2 + v_y{}^2} \right)$.

The angle of the projectile can be found at any point by taking the inverse tangent of $\dfrac{v_y}{v_x}$ at that point.

Projectile motion is dependent on having no other forces present other than gravity. So the analysis is different during launch, during the landing, under the effects of friction, or for an object with its own source of thrust. For this reason, physicists tend to reserve the use of the word *projectile* for free-fall conditions only.

 **THINGS TO THINK ABOUT**

Students new to physics will sometimes set the final velocity to be zero in a projectile problem. After all, doesn't the projectile come to rest after it strikes the ground? The problem with this is that as soon as the projectile begins to touch the ground, the free-body diagram changes and thus the acceleration is no longer the same. Recall that the equations of motion we are using work only under the conditions of *constant* acceleration.

**EXAMPLE** A projectile is launched at a 25-degree angle at a speed of 45 m/s from ground level.

a. How high does the projectile go?

b. What is the projectile's speed and angle of flight (velocity vector) 3.0 seconds into flight?

c. What is the range (horizontal distance upon impact) of the particle?

**SOLUTION**

a. At the highest point, $v_y = 0$. Find the time by using:

$$v_y = v_{yi} - gt$$

$$0 = (45\sin 25°) - 9.8t$$

$$t = 1.94 \text{ s}$$

Now find the vertical displacement by using:

$$\Delta y = v_{yi}t - \frac{1}{2}gt^2$$

$$\Delta y = (45\sin 25°)(1.94) - \frac{1}{2}(9.8)(1.94)^2 = 18.5 \text{ m}$$

b. First find $v_y$ at the given time:

$$v_y = v_{yi} - gt = (45\sin 25°) - 9.8(3) = -10.4 \text{ m/s}$$

The negative sign indicates that the projectile is moving downward. Now find $v_x$, which never changes:

$$v_x = 45\cos 25° = 40.8 \text{ m/s}$$

Calculate the speed of the projectile:

$$\text{Speed} = \sqrt{v_x^2 + v_y^2} = \sqrt{40.8^2 + (-10.4)^2} = 42.1 \text{ m/s}$$

Now use the inverse tangent to calculate the angle of the projectile:

$$\theta = \tan^{-1}\left(\frac{v_y}{v_x}\right) = \tan^{-1}\left(\frac{-10.4}{42.1}\right) = -13.9°$$

c. Double the time found from part (a) to determine the time in flight:

$$t = 3.88 \text{ s}$$

Now use the horizontal displacement formula and the velocity in the x-direction found in part (b):

$$\Delta x = v_{xi}t = 40.8(3.88) = 158 \text{ m}$$

# 4.4 Orbital Motion

Orbital motion is circular motion in which the centripetal force is supplied by the gravitational force. Since orbital motion takes place far from the surface of a planet, Newton's universal gravitation formula should be used:

$$F_{\text{net}} = F_{\text{g}} = ma_{\text{c}} = \frac{mv^2}{r}$$

$$G\frac{Mm}{r^2} = \frac{mv^2}{r}$$

In this equation, $M$ is the central mass (e.g., planet) and $m$ is the mass of the orbiter. Note that the $r$ has two identical values in this case. It is both the distance between the center of the planet and the orbiter and is also the radius of the circle. Canceling like terms gives:

$$\frac{GM}{r} = v^2$$

This is the algebraic explanation for some of Kepler's observations in 1619 regarding planetary motion: the orbital velocity ($v$) does not depend on the mass of the orbiter and depends only on distance ($r$) from a given central object. Thus, the planets move more slowly the farther they are from the sun. Of course, this holds for all objects in orbit around the same central object: satellites in low Earth orbit must travel much more quickly to maintain their orbit than satellites at higher altitudes. Another of Kepler's laws of planetary motion can be found algebraically as well by rewriting the orbital velocity ($v$) in terms of the orbital period ($T$), which is the time to complete one revolution:

$$v = \frac{2\pi r}{T}$$

Substituting this expression in the orbital velocity equation above and rearranging gives:

$$\frac{T^2}{r^3} = \frac{4\pi^2}{GM}$$

With Newton's help, we can now explain and give physical meaning to Kepler's observation that all the planets have the same $T^2/r^3$ ratio. In fact, the ratio is uniquely determined by the mass of the central object. This is how astronomers make the most accurate determination of the masses of distant objects: they measure the period and distance of an object in orbit *around* the unknown mass and use this simple formula to solve for $M$.

**THINGS TO THINK ABOUT**

When Henry Cavendish first measured the value for the universal gravitational constant, $G$, in 1797, it was known as the experiment that "weighed the Earth."

**THINGS TO THINK ABOUT**

Newton himself pointed out long ago that orbital motion is simply an extreme case of projectile motion. If a projectile is launched fast enough, as it falls to Earth the planet's own curvature could match the curvature of the projectile: the projectile's horizontal speed would then be its tangential circular speed (also known as the object orbital speed). The feeling of weightlessness that astronauts experience in orbit is not due to the lack of gravity (obviously; no gravity would mean no circular/orbital motion) but, rather, because they are constantly in free fall. They are no more weightless than you are when you trip and fall and have yet to hit the ground!

**EXAMPLE** Use the mass of the Earth and the distance to the moon to determine the orbital speed the moon must have to maintain its orbit.

**SOLUTION**

$$M_{Earth} = 6.0 \times 10^{24} \text{ kg}$$

Earth-moon distance $= r = 3.8 \times 10^8$ m

$$v^2 = \frac{GM}{r}$$

$$v^2 = \frac{\left(6.67 \times 10^{-11}\right)\left(6.0 \times 10^{24}\right)}{3.8 \times 10^8}$$

$$v = 1{,}020 \text{ m/s}$$

# Chapter Review Exercises

Your rowing speed in calm waters is at 2.6 m/s ($v_0$). The river you must cross has a current of 1.8 m/s ($v_c$) heading toward the east from the west.

1. You angle your boat to cross the river straight across (due north). However, the current pushes you downstream and your resultant ($R$) velocity has you landing downstream on the other side.

   a. Draw a graphical tip-to-tail addition of $v_0$ and $v_c$, labeling the resultant $R$.

   b. Will the resulting speed be greater or smaller than the rate of 2.6 m/s that you are rowing?

   c. Find the angle and magnitude of the resultant. Indicate the angle on your drawing.

2. This time, you angle your boat upstream so that your resultant will take you straight across the river.

   a. Draw a graphical tip-to-tail addition of $v_0$ and $v_c$, labeling the resultant $R$.

   b. Will the resulting speed be greater or smaller than the rate of 2.6 m/s that you are rowing?

   c. Find the angle of your $v_0$. Indicate the angle on your drawing.

3. You fire a cannonball of mass 10.0 kg out of a cannon. You measure the velocity of the cannonball as it comes out of the cannon to be 15.0 m/s at an angle of 70.0 degrees to the horizontal. You are on a level field on Earth, and you may ignore the effects of friction.

   a. How long is the cannonball in flight?

   b. How far along the ground (the range) does the cannonball travel?

   c. How many seconds does the cannonball take to reach its highest point?

d.  Fill in the chart with the components of the velocity and displacement vectors at the indicated times.

| Time | x-Position | y-Position | $v_x$ Component | $v_y$ Component |
|---|---|---|---|---|
| Initial | | | | |
| At cannonball's highest point | | | | |
| As cannonball hits the ground | | | | |

e.  What is the speed and angle of the cannonball's velocity as it hits the ground?

4.  Using only the known orbital distance of the Earth to the sun ($1.50 \times 10^8$ km) and the period of the Earth's revolution about the sun (one year), determine the mass of the sun.

5.  You are swinging a tiny teacup of water at the end of a short string at constant speed in a circle.

a.  Briefly explain why there is acceleration even though the speed is constant.

b.  If the string breaks when the tension is greater than 150 newtons, how fast can you swing the teacup horizontally before it breaks (mass of teacup = 0.50 kg, length of string = 0.25 m)?

c.  Briefly explain why the tension in a circle swung horizontally is constant, whereas the tension in a circle swung vertically is not.

d.  At what point in the vertical circle is the string most likely to break? Determine the maximum speed using the values from part (b).

# ANGULAR DYNAMICS AND KINEMATICS

## WHAT YOU WILL LEARN

- All of Newton's laws are true for rotational motion just as they are for linear motion.

- How the kinematics of rotating objects works.

- What a torque is and how it affects rotation.

- The role of the moment of inertia in rotational dynamics.

- The role of the center of mass in determining if an object is stable or unstable.

- How to solve static situations using forces and torques.

| LESSONS IN CHAPTER 5 | |
|---|---|
| • Rotational Kinematics | • Center of Mass |
| • Rotational Dynamics | • Static Problems |

## PARALLELS BETWEEN LINEAR AND ROTATIONAL PHYSICS

In addition to linear dynamics and the resulting linear kinematics, physics is also concerned with rotating objects. In his original work, Newton actually addressed both linear and rotational motion at the same time in his three laws of motion.

**THINGS TO THINK ABOUT**

Note that a revolution is when an object is undergoing circular motion about an external point, whereas a rotation is when the object itself is spinning about an internal axis.

All of the rotational kinematics and almost all of the rotational dynamics can be understood via analogy with the linear versions we have already discussed.

1.  When tracking rotations, use radians of angular displacement rather than meters of linear displacement.

2.  When a change in rotational motion is required, use torques rather than simple forces.

3.  When inertia is required in the rotational world, use the moment of inertia rather than the simple mass of an object.

# 5.1 Rotational Kinematics

The proper unit of angular measurement is the radian. Note that the radian, which is abbreviated as rad, is not a true unit in the sense that it is a dimensionless ratio. However, it is useful to track the radian as if it were a unit. The definition of the radian measure of an angle ($\theta$) is the ratio of the arc length ($s$) of the circle defined by that central angle divided by the radius of the circle ($r$). As shown in Figure 5.1, the radian measure of an angle is found using $\theta = \dfrac{s}{r}$.

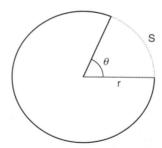

**FIGURE 5.1** The radian measure of an angle.

When translating among angular units, the exact conversions are as follows:

1 rotation or revolution = 360 degrees = $2\pi$ radians

For rotational kinematics, simply replace all linear variables in the linear equations of motion with their rotational counterparts (typically symbolized with Greek letters). This is shown in Table 5.1.

### TABLE 5.1  VARIABLES IN LINEAR EQUATIONS AND IN ROTATIONAL KINEMATICS

| Linear | Angular/Rotational |
|---|---|
| $x$ for position in m | $\theta$ (theta) for angle in radians (rad) |
| $v$ for velocity in m/s | $\omega$ (omega) for angular speed in rad/s |
| $a$ for acceleration in m/s² | $\alpha$ (alpha) for angular acceleration in (rad/s²) |

The following shows the linear equation for position on the left and the corresponding rotational equation for angle on the right:

$$\Delta x = v_0 t + \frac{1}{2}at^2 \qquad \Delta\theta = \omega_0 t + \frac{1}{2}\alpha t^2$$

**EXAMPLE**  An old-school turntable can take a record from stationary to 45 rpm in about half a second. What is the average acceleration of the turntable (in rad/s²)?

### SOLUTION

First let's change the units of rpm (rotations per minute or rot/min) into the standard units of radians/second:

$$\left(\frac{45 \text{ rot}}{\text{min}}\right) \times \left(\frac{2\pi \text{ rad}}{\text{rot}}\right) \times \left(\frac{1 \text{ min}}{60 \text{ s}}\right) = 4.7 \text{ rad/s} = \omega_f$$

You are given:

$$\omega_i = 0 \text{ rad/s}$$
$$t = 0.5 \text{ s}$$

Use the rotational equation analogous to the kinematic equation $x_f = x_i + at$, substitute the known values, and solve for alpha:

$$\omega_f = \omega_i + \alpha t$$
$$(4.7 \text{ rad/s}) = 0 \text{ rad/s} + (0.5 \text{ s})\alpha$$
$$\alpha = 9.4 \text{ rad/s}^2$$

# 5.2 Rotational Dynamics

When rotating, everything is relative to the axis of rotation. In fact, most rotational quantities are undefined unless an axis of rotation is specified. Both the efficacy of a force in causing rotation and the effect of a mass on the rotational inertia are related to how far from the application of the force or the distribution of the mass is from the axis of rotation.

## TORQUE

**Torque** ($\tau$, "tau") is the measure of how effective a force is at causing a rotation (see Figure 5.2). It is the combination of the force applied to an object and the distance from the axis of rotation to the point of application of the force. Torque is a vector quantity and has units of N • m. Only the portion of the force that is tangent to the circular motion at the point of contact is responsible for changing rotational motion. This is sometimes written as $F_\perp$ since the component of the force tangent to the circular motion is perpendicular to the radius of the motion. To find this component, simply multiply the magnitude of force times the sine of the angle formed by the $r$ and $F$ vectors.

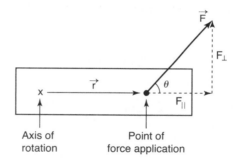

**FIGURE 5.2** Torque.

Formally, the torque is a cross product, which entails a bit of additional mathematics for vectors that is usually beyond the scope of an algebra-based course:

$$\vec{\tau} = \vec{r} \times \vec{F}$$

However, in most first-year physics classes, a more informal approach is used for directions involving rotations. Torques and the other rotational vector quantities, such angular displacement and angular velocity, are denoted as either clockwise or counterclockwise. Even if the torque is not actually causing a rotation, its direction is defined based on the motion the force would produce. The magnitude of the torque can now be calculated using the following expressions:

$$\tau = |\vec{r}|F_\perp = r_\perp|\vec{F}| = |\vec{r}||\vec{F}|\sin\theta$$

 **THINGS TO THINK ABOUT**

*A note about signs*

Formally, there is a direction in space that is associated with the cross product of two vectors using something called the right-hand rule. Because of this, the sign convention for all rotational direction is for clockwise to be positive and counterclockwise to be negative. If this seems counterintuitive, consider the Cartesian coordinate system shown in Figure 5.3. Note how the standard angle is defined and how mathematicians number the quadrants.

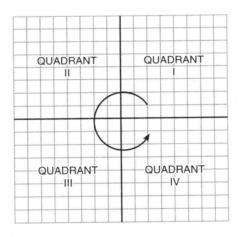

**FIGURE 5.3** Cartesian coordinate system.

All of the various ways of writing what the force is being multiplied by in the torque equation is sometimes called the lever arm of the situation (see below for a classic lever problem). In situations with more than one torque being applied, determine the direction of each torque independently with its magnitude. Then, add up all the individual torques with their proper signs. The sign of the sum will indicate the direction of the net torque.

**EXAMPLE**    Calculate the net torque due to the three forces (each is 4.5 N strong) on the disk
(rotated about its center) pictured below. The disk's radius is 0.84 m.

**SOLUTION**

The force on the left that is pointed down causes a counterclockwise (CCW)
torque. The force pointed down on the right causes a clockwise (CW) torque.
The force pointed to the right toward the top also causes a CW torque.

Since all forces are perpendicular to the radius at their location and they all
have the same magnitude, the magnitude of torque is the same for each:

$$\tau = |\vec{r}|F_{\perp} = (0.84 \text{ m})(4.5 \text{ N}) = 3.78 \text{ N} \cdot \text{m}$$

Sum up the torques in the order described above. Let the CCW torque be
positive and the CW torques be negative:

$$+3.78 \text{ N} \cdot \text{m} - 3.78 \text{ N} \cdot \text{m} - 3.78 \text{ N} \cdot \text{m} = -3.78 \text{ N} \cdot \text{m}$$

The negative sign indicates a CW direction for net torque

# MOMENT OF INERTIA

The rotational inertia of an object about a specified axis of rotation is known as its
moment of inertia or rotational inertia ($I$). Like mass, rotational inertia is a scalar quantity
and represents the object's resistance to changes in rotational motion. Unlike mass,
though, its units are kg • m$^2$. The formal way of finding the moment of inertia of an actual
object about a specific axis is generally beyond the scope of an algebra-based physics
class. However, in the special case of a simple object consisting of specified masses
($m$) at particular distances from the axis of rotation ($r$), you can calculate the moment of
inertia with the following equation:

$$I = \Sigma mr^2$$

Note that the moment of inertia is not technically defined for an object in general; it is
defined only for a given object rotating about a *specific* axis. If you look up, for example,
the moment of inertia of a sphere and find that it is $\frac{2}{5}mr^2$ without an axis specified, the
assumption is that the object is rotating about its center.

Take an idealized dumbbell used in weightlifting.

If each side of the dumbbell is 10 kg and the distance between the sides is 0.20 meters, calculate the moment of inertia for the dumbbell if it is rotated:

a.  About its midpoint

b.  About one of the weights

(Ignore the mass of the central bar.)

**SOLUTION**

a.  About the midpoint:

$$I = \Sigma mr^2 = m_1 r_1{}^2 + m_2 r_2{}^2$$
$$I = (10 \text{ kg})(0.1 \text{ m})^2 + (10 \text{ kg})(0.1 \text{ m})^2 = 0.2 \text{ kg} \cdot \text{m}^2$$

b.  About one end:

$$I = \Sigma mr^2 = m_1 r_1{}^2 + m_2 r_2{}^2$$
$$I = (10 \text{ kg})(0 \text{ m})^2 + (10 \text{ kg})(0.2 \text{ m})^2 = 0.4 \text{ kg} \cdot \text{m}^2$$

Try taking any symmetrical rod (it doesn't even have to have weights). Try rotating it quickly about its midpoint and then again about one of its ends. You will find it *is* more difficult to change the rod's rotation when rotating about its end as indicated in the example by the higher moment of inertia calculation!

# NEWTON'S SECOND LAW FOR ROTATION

Newton's second law for rotation states that net torque on an object equals the object's moment of inertia times its angular acceleration.

$$\vec{\tau}_{net} = I\vec{\alpha}$$

You can use the equation listed above to find the net torque. If the object is not rotating or if it is rotating at constant angular speed (omega, ω), the angular acceleration (alpha, α) in the equation above is zero.

**EXAMPLE** The classic torque problem is that of a seesaw.

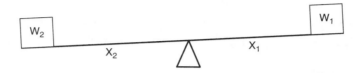

Find a general relationship between the variables above when the seesaw is balanced.

### SOLUTION

The seesaw can be made to balance about the fulcrum (the axis of rotation) even if the weights ($W_1$ and $W_2$) are different by adjusting their distances ($X_1$ and $X_2$). The problem is further simplified if the seesaw is assumed to be close to horizontal. This makes the angle between the gravitational force vector and the radial vector perpendicular, resulting in the sine function equaling 1 (sin90°). Weight $W_2$ causes a counterclockwise rotation, whereas weight $W_1$ causes a clockwise rotation:

$$\vec{\tau}_{net} = I\vec{\alpha}$$

$$W_2 X_2 - W_1 X_1 = I\vec{\alpha}$$

Since you are told the seesaw is balanced, $\alpha = 0$:

$$W_2 X_2 - W_1 X_1 = 0$$

$$W_2 X_2 = W_1 X_1$$

## 5.3 Center of Mass

If you balance an object on a point or hang the object with a string, the object's center of mass will be directly in line with that contact point. The object is in equilibrium (with respect to torques). The center of mass is also the idealized location of an object. When we draw a free-body diagram or use kinematics to describe an object's location, velocity, or acceleration, we are actually describing the object as if it was a point of mass concentrated at the object's center of mass. No internal forces can change the center of mass of an object. It is the center of mass that is obeying Newton's laws. For example, in Figure 5.4 one can see the actual flight of a hammer through the air can be quite complicated. However, the center of mass of the hammer follows a perfect parabola as in projectile motion.

**FIGURE 5.4** Compare center of mass trajectory to the parabolic path of a projectile.

Sometimes called the center of gravity, the center of mass and the center of gravity are the same for most objects. Some large objects, like moons or planets, may have slightly different centers of gravity as compared with their center of mass. Specifically, this occurs if they are so big that the gravitational field they are in cannot be approximated as constant.

The center of mass of an object can be determined in two ways:

- For distributions of masses, you must use calculus.

- For a distribution of point masses, you can take a weighted average along each axis.

Use the following equations for a distribution of point masses:

$$m_{total} = \sum m_i$$

$$\vec{r}_{center\ of\ mass} = \frac{\sum m_i \vec{r}_i}{m_{total}}$$

Figure 5.5 shows the relationship between center of gravity (CG), the pull of gravity, and torque.

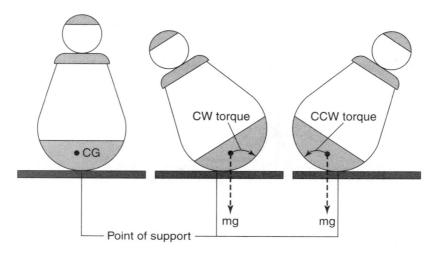

Use the point of contact as the axis of rotation.

**FIGURE 5.5** The rotational stability of an object with a low center of gravity.

## STABILITY

If a small, sideways, outside force (a *perturbation*) raises the center of mass of an object, it will cause a gravitational torque to rise in the opposite direction of the tipping action, causing the object to return to its original position. Since the object returns to its original position under these conditions, it is in a stable equilibrium.

If a perturbation causes the center of mass to lower, the object will continue to tip over, even if the outside force is slight. This is known as an unstable equilibrium.

# 5.4 Static Problems

A static situation is one in which there is no linear motion as well as no rotational motion. Dynamically, this means that acceleration ($a$) and rotational acceleration ($\alpha$) are both zero. Thus, the starting point for all static problems is to write Newton's second law three times:

$$F_{net,x} = \Sigma F_x = 0$$
$$F_{net,y} = \Sigma F_y = 0$$
$$\vec{\tau}_{net} = 0$$

The two linear equations given above are straightforward. Frequently in this style of problem, though, the two equations will be insufficient for solving the problem. In other words, there will be more than two unknowns. The freedom in defining one's own axis in the torque equation is the key to these problems. Since the object is not rotating at all, it is not rotating about *any* possible axis. Simply choose to consider the torques about the point of application of some unknown forces. Those particular forces will then be eliminated in the torque equation since their lever arms will be zero. Note that the point of contact for the gravitational force on an object is its center of mass. Often, it is not even necessary to solve the force equations. So start these problems by analyzing the net torque.

EXAMPLE   Determine the tension in the line supporting the 175 N pole (whose center of mass is indicated with an **X** that is 0.30 meters from the wall). The sign has a mass of 5 kg and is 0.15 meters from the wall. The pole is 0.80 meters long.

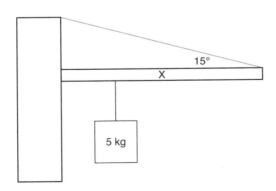

## SOLUTION

The point at which the pole meets the wall involves several unknown forces. In fact, the exact mechanism of attachment is not even specified. Generally speaking, the unknown contact forces at this point (represented by $C$) will have both a horizontal and a vertical component, leading to the unsolvable system of linear equations. The other forces are the tension in the line ($T$), the weight of the pole itself (175 N), and the weight of the sign (5 kg):

$$C_x - T\cos15° = 0$$
$$C_y + T\sin15° - 175 - 5(9.8) = 0$$

However, the tension can be found from the torque equation alone by considering the net torque about this same contact point:

$$\tau_C + \tau_T + \tau_{sign} + \tau_{pole} = 0$$

$$C(0) + T(0.80)\sin15° - 5(9.8)(0.15)\sin90° - 175(0.3)\sin90° = 0$$

Solve for $T$:

$$T = 289 \text{ N}$$

# Chapter Review Exercises

1. Opening (or closing) a door is a torque problem. Give two examples of how force can be applied to open a door and yet have no resulting torque.

2. Place a pencil flat on a table. Now apply two equal forces in opposite directions such that the net force is zero and yet there is a net torque that causes the pencil to rotate. Explain.

3. If a high diver can execute 4.5 flips in 1.5 seconds during descent, estimate the required torque off of the diving board. Assume the diver's push off the board was about 0.15 seconds in duration, and use the fact that divers can vary their moment of inertia during the dive from a high of over 12 kg • m² to less than 6 kg • m².

4.

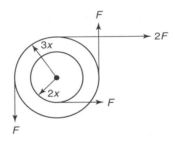

a. In terms of $F$ and $R$, find $\tau_{net}$ on the two-wheel system shown above. The wheels are fixed to each other and are free to rotate together about the center of both.

b. If $F = 24.5$ N and $R = 1.5$ meters, find the angular acceleration, $\alpha$, of the system. You can assume the moment of inertia of the system is 16.0 kg • m².

c. If the system starts from rest, what will be its final angular velocity, $\omega_f$, after 8.0 seconds?

d. In those same 8.0 seconds, what is the total angular displacement of the system?

5. The horizontal uniform rod shown below has a length of 0.60 m and a mass of 2.0 kg. The left end of the rod is attached to a vertical support by a frictionless hinge that allows the rod to swing up or down. The right end of the rod is supported by a line that makes an angle of 30° with the rod. A 0.50 kg block is also attached to the right end of the rod.

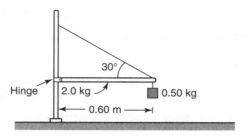

a. Calculate the tension in the line.

b. Find the force that the hinge exerts on the rod.

# CONSERVATION LAWS

## WHAT YOU WILL LEARN

- Symmetries in nature underlie all conservation laws.

- How linear momentum is conserved during impacts.

- The role of conservation of angular momentum in rotational motion.

- The role of the moment of inertia in rotational dynamics.

- How energy is transformed and used.

- The role of energy in bonds and in mass.

| LESSONS IN CHAPTER 6 | |
| --- | --- |
| • Why Are There Conservation Laws? | • Conservation of Energy |
| • Conservation of Linear Momentum | • More on Energy |
| • Conservation of Angular Momentum | |

## 6.1 Why Are There Conservation Laws?

In science, a conserved quantity refers to an amount or a measure of some attribute of a system that does not change upon internal interactions and changes. For example, electric charge is a conserved quantity. Charge can be transferred, destroyed, and created, but the net amount of charge in a closed system must always total the same amount. If charge is transferred into the system from an external source, the gain of charge within the system must be the same as the loss of charge from that external source. (Indeed, one can simply imagine drawing a larger box and defining the system under consideration to include the source of the transfer as well, and charge must be conserved in that larger system.) Why are some quantities conserved, and which ones are conserved? In 1918, Emmy Noether answered these questions with what is considered by many to be one of the most profound truths in science: Behind every conserved quantity is a deep symmetry in nature (see Table 6.1). This connection is known as **Noether's Theorem**. The conserved quantity and the physical variable that demonstrate the symmetry form a conjugate pair, which is the very same pair of variables in the Heisenberg uncertainty principle in quantum mechanics (see Chapter 15). In this chapter, we will be examining the three most important conserved quantities.

**TABLE 6.1  SYMMETRY, CONSERVED QUANTITIES, AND CONJUGATE VARIABLES**

| Symmetry: The laws of physics do not change with respect to . . . | Conserved quantity: The total amount of this quantity does not change in an isolated system | Conjugate variables: The product of these two variables obeys the Heisenberg uncertainty principle |
|---|---|---|
| Location (spatial translation) | Linear momentum ($p$) | $x, p$ |
| Orientation (rotation) | Angular momentum ($L$) | $\theta, L$ |
| Time | Energy ($E$) | $t, E$ |

# DEFINING THE RIGHT SYSTEM

As explained above, whether a quantity is conserved depends on how the system is defined. Consider the collision of two cars. If your system is one of the two cars, its momentum is certainly not conserved since that car is not an isolated system; forces external to the system are present during the collision. However, if your system includes both cars, the total momentum is conserved during the collision since the forces of collision are internal to the system. This business of choosing the elements to include in your system for a given problem can be tricky for the beginning student. However, if an object involved in the problem has unknown attributes (mass, initial velocity, etc.), it is probably wise to make that object external to the system. See the first few problems at the end of the chapter to practice choosing the best system to track.

# 6.2 Conservation of Linear Momentum

Linear momentum is inertia in motion. One simple concept to capture linear momentum is to think about how difficult it is to stop the forward motion of an object. An object with small mass could be more difficult to stop than a much larger mass if the small mass was traveling at a much higher velocity. The momentum ($p$) of an object is simply the product of its mass ($m$) and velocity ($v$):

$$\vec{p} = m\vec{v}$$

Note that momentum is a vector quantity and gets its direction uniquely from the velocity. Momentum does not have its own defined units but is simply kg • m/s in the metric system. The net momentum of a system is simply the vector sum of all the momenta of the individual particles in the system:

$$\vec{p}_{net} = \vec{p}_1 + \vec{p}_2 + \vec{p}_3 + ... = \Sigma \vec{p}_i$$

Impulse is the mechanism by which momentum is transferred from one object to another. Impulse is the amount of external force times the amount of time that force is applied to the object:

$$\vec{F}\Delta t = \Delta \vec{p} = \vec{p}_f - \vec{p}_i$$

Here in the impulse equation (sometimes called the impulse momentum theorem), we have the secret to almost all crash safety devices (think seat belt, air bag, crumple zones, breakaway highway poles, etc.). By increasing the time taken ($\Delta t$) to accomplish the same change in momentum ($\Delta \vec{p}$), the force required is made to be less. Take, for instance, a car crash where the driver goes from 55 mph to 0 mph. It's much better for the driver's body to make that change in 0.3 seconds rather than in 0.03 seconds!

If no external force is applied to a system, the total linear momentum is conserved:

$$\Sigma \vec{p}_i = \Sigma \vec{p}_f$$

**EXAMPLE** A bowling ball that is five times heavier than a stationary billiard ball strikes the stationary ball with a velocity of 8 m/s. After the impact, both balls roll in the same direction, the bowling ball continuing with only 4 m/s in velocity. What is the velocity of the billiard ball?

Before Collision          After Collision

8 m/s                     4 m/s          ?

**SOLUTION**

Drawing before and after pictures of the collision is helpful in solving conservation of momentum problems. Simply add up all the momenta in the before picture and compare to the sum of the momenta in the after picture. Let $x$ be the unknown mass of the billiard ball:

$$\text{Before: } m_1 v_1 + m_2 v_2 = (5x)(8) + (x)(0) = 40x$$

$$\text{After: } m_1 v_1 + m_2 v_2 = (5x)(4) + (x)v = x(20+ v)$$

Set these two quantities equal to each other (conserving momentum):

$$40x = x(20+v)$$

Solve for $v$:

$$v = 20 \text{ m/s}$$

# 6.3 Conservation of Angular Momentum

Angular momentum is inertia in rotation. Picture how difficult it is to stop something from spinning. The angular momentum $(\vec{L})$ is simply the product of the moment of inertia $(I)$ and angular velocity $(\vec{\omega})$, both previously defined in Chapter 4. Angular momentum is a vector with its direction taken from the direction of angular velocity and having units of kg • m$^2$/s$^2$:

$$\vec{L} = I\vec{\omega}$$

A torque applied for a certain interval of time induces a change in angular momentum:

$$\vec{\tau}\Delta t = \Delta \vec{L}$$

(Compare this to the impulse-momentum theorem for linear motion.)

If no external torque is applied to a system, the total angular momentum is conserved:

$$\Sigma \vec{L}_i = \Sigma \vec{L}_f$$

**EXAMPLE** As a certain ice skater draws her arms in tightly to her spinning body, she decreases her momentum of inertia by a factor of 3. If she was spinning at the rate of half a rotation per second, what is her new rate of spin?

**SOLUTION**

Since the ice skater is not receiving any external torques during her spin, she is an isolated system. Therefore, her angular momentum must be conserved:

$$I_i \omega_i = I_f \omega_f$$

Since $I_f = \dfrac{1}{3} I_i$ and $\omega_i = 0.5$ rot/s:

$$I_i(0.5) = \frac{1}{3} I_i \omega_f$$

Solve for $\omega_f$:

$$\omega_f = 1.5 \text{ rot/s}$$

# 6.4 Conservation of Energy

Energy is the most important concept in science but also an extremely abstract idea. Energy is a scalar quantity that represents an object's ability to impose change on its environment or itself. Although there are several different types of energy, the most essential form is that of **kinetic energy**. In equations, kinetic energy (KE) often appears as $K$ and is defined as:

$$K = \frac{1}{2} m v^2$$

Indeed, if we ignore any internal structure of an object, kinetic energy is the only type of energy an object in isolation can have. If the system under consideration is isolated, the kinetic energy changes when external forces do work ($W$) on it:

$$W_{net} = \Delta K$$

The work done by each individual force is determined by taking the dot product of force and the displacement of the object:

$$W = \vec{d} \cdot \vec{F} = |\vec{d}| F_{//} = d_{//} \vec{F} = |\vec{d}||\vec{F}| \cos\theta$$

# UNITS OF WORK AND ENERGY

By virtue of the kinetic energy formula, energy should have units of kg • m$^2$/s$^2$. However, the work done in increasing or in decreasing the energy has units of N • m. A little dimensional analysis proves that these are the same units. The metric unit for work and energy is the joule, which is equal to the following combination of base units:

$$N \bullet m = \left(\frac{kg \bullet m}{s^2}\right)(m) = \frac{kg \bullet m^2}{s^2} = joule \ (J)$$

Note that the dot product of two vectors involves the components of force and displacement working in the same direction. As such, work is a scalar (as are all energies in general) and has no direction associated with it. If the work is positive, that force is adding energy to the system; if it is negative, that force is taking energy away from the system.

**EXAMPLE** A student is pushing a 0.5 kg book across a tabletop at a constant velocity by using a force of 2.5 N. After pushing it for 1 meter, the student asks the physics teacher, "Why isn't the book speeding up if I'm applying a force over a distance?"

### SOLUTION

The student is doing positive work to the book:

$$|\vec{d}||\vec{F}|\cos\theta = (1)(2.5)(1) = 2.5 \ joules$$

However, the book is not accelerating and is therefore not gaining any kinetic energy. According to Newton's second law, the net force must be zero since the book is not accelerating. The frictional force between the table and the book must be canceling out the student's push. The work done by the frictional force is at a 180-degree angle to the displacement. Calculate the work done by friction:

$$|\vec{d}||\vec{F}|\cos\theta = (1)(2.5)(-1) = -2.5 \ joules$$

Calculate the net or total work:

$$+2.5 \ J + -2.5 \ J = 0 \ J$$

There is no change in kinetic energy.

Although this simple version of energy is consistent, it is not yet in the form of a conservation law. The increase in an object's kinetic energy does not come at the corresponding loss of kinetic energy elsewhere in the universe. Consider a falling rock.

The increasing kinetic energy of the rock as it falls does not come at the expense of Earth's own kinetic energy, despite the fact that Earth's gravitational force is doing the positive work on the rock. To get to the conserved energies, we must first consider that the work done by *individual* forces comes in two different flavors: work done by conservative forces and work done by nonconservative forces. **Conservative forces** do work that is path independent and have potential energies associated with them. **Nonconservative forces** do work that is path dependent and thus cannot have potential energies associated with them. There are only a limited number of conservative forces:

# CONSERVATIVE FORCES

- Gravity

- Elastic Forces (obeying Hooke's law)

- Electric and magnetic forces

- The strong and weak nuclear forces

We must now expand our system to include the object that is applying the conservative force to the original object. The potential energy (PE) associated with each of these internal forces is easily found by calculating the work done by that particular force and bringing it to the energy side of the equation. Divide the net work into the two different types of work, nonconservative work ($W_{nc}$) and conservative work ($W_c$):

$$W_{net} = \Delta K$$

$$W_{nc} + W_c = \Delta K$$

$$W_{nc} = \Delta K - W_c$$

Enlarge the system to include the object applying the conservative force and defining the change in potential energy ($\Delta U$) as negative work done by the conservative force:

$$W_{nc} = \Delta K + \Delta U$$

What happened to the negative sign for the work done by conservative forces when changing systems to consider it as stored potential energy? The negative work done by conservative forces that was previously considered external to the system is now considered potential energy held within the system. The system can get those joules of potential energy right back as kinetic energy—they are no longer lost, they are stored! Note talking about potential energy external to a system does not make sense.

Potential energies are not owned by a single object. Rather, **potential energy** is stored in the system in the relationship between two objects. When you lift a book higher above the ground, there is greater potential energy in the Earth-book system than when it was lower, but the book by itself does not have any potential energy.

The following are the two most common forms of potential energy:

$$U_g = mgh$$

This first form is the potential energy stored in the Earth-mass ($m$) system relative to an arbitrary point $h = 0$.

$$U_s = \frac{1}{2}kx^2$$

This second form is the potential energy stored in the spring connecting two objects on either end. The spring is characterized by the spring constant $k$ (in N/m) and the amount the spring has been stretched or compressed away from its relaxed length, $x$ (in meters).

 **THINGS TO THINK ABOUT**

*Gravitational potential energy*

If a system is such that it is fair to use the simplified form of gravity ($F_g = mg$), then a simplified form of gravitational potential energy is also fair to use:

$\Delta U_g = mg\Delta h$, where $h$ is height. However, if the universal form of gravity is being used $\left( F_g = \dfrac{Gm_1m_2}{r^2} \right)$, the universal gravitational potential energy form must also be used, $\left( U_g = -\dfrac{Gm_1m_2}{r^2} \right)$.

This combination of the kinetic and potential energies into a single **mechanical energy** (ME) is a conserved quantity. If no work is done by nonconservative forces in an isolated system:

$$(K + U)_{\text{initial}} = (K + U)_{\text{final}}$$

**EXAMPLE** A toy dart gun shoots lightweight felt darts ($m = 0.025$ kg) by launching the darts from a compressed spring ($k = 12$ N/m). If a dart is to be launched at 1.5 m/s, how far must the spring be compressed?

**SOLUTION**

Since there are no external forces and no nonconservative forces on the spring-dart system, we can use conservation of energy:

$$(K + U)_{initial} = (K + U)_{final}$$

Initially the dart is not moving, but the spring is fully compressed a distance $x$. In the final state, the dart is moving, but the spring is no longer compressed:

$$0 + \frac{1}{2}kx^2 = \frac{1}{2}mv^2 + 0$$

Solve for $x$:

$$x = \sqrt{\frac{m}{k}}v$$

$$x = \sqrt{\frac{0.025}{12}} \times 1.5 = 0.068 \text{ m}$$

**EXAMPLE** **Power** is defined to be the rate at which work is done (or energy is transferred). Its metric unit is the watt (W). Rewrite the watt in terms of the base SI units meter, second, and kilogram.

**SOLUTION**

Rates are made by dividing the quantity (work) by the time that quantity takes:

$$\frac{\text{work}}{\text{time}}$$

These two quantities have units of joules and seconds:

$$\text{watt} = \frac{\text{joule}}{\text{second}} = \frac{\text{Newton} \cdot \text{meter}}{\text{second}} = \frac{\left(\frac{\text{kg} \cdot \text{m}}{\text{s}^2}\right)(\text{m})}{\text{s}} = \frac{\text{kg} \cdot \text{m}^2}{\text{s}^3}$$

Whew! No wonder everyone talks about power in watts!

# WHICH WORK? WHAT'S CONSERVED?

Table 6.2 summarizes the various types of work.

**TABLE 6.2  A SUMMARY OF THE VARIOUS TYPES OF WORK**

| | |
|---|---|
| Work done by a single force | $W = \vec{F} \bullet \vec{d} = Fd\cos\theta$ |
| Work done by **all** the forces acting on an object | $W_{net} = \Delta KE$<br>where KE = kinetic energy = $\frac{1}{2}mv^2$ |
| Work done by only nonconservative forces | $W_{nc} = \Delta ME$<br>where ME = mechanical energy = KE + PE,<br>where KE = kinetic energy,<br>where PE = potential energy |

In the equation ME = KE + PE, there is one term for each conservative force present.

There are three common forms of potential energy (PE):

- Gravitational = $mgh$ (near the Earth's surface) and $-\dfrac{GMm}{r}$ (universal)
- Elastic = $\dfrac{1}{2}kx^2$
- Electric = $qV$

# 6.5  More on Energy

## INTERNAL ENERGY

In addition to their macroscopic kinetic energies and potential energies, objects that have internal structure can also store energy internally. Oftentimes these internal energies are ignored—the entire field of thermodynamics (see Chapter 13) is the tracking of these internal energies. It is worthy to note, however, that this internal energy is not a new form of energy. It is simply microscopic kinetic energy (something we use temperature as a proxy for) and microscopic potential energies (e.g., chemical potential energy). To remove our former stipulation about mechanical energy being conserved only in the absence of nonconservative forces, we must add this internal energy to the mix. To see true energy conservation, one must track all three forms of energy:

Total Energy = Kinetic Energy + Potential Energy + Internal Energies

In an isolated system, this total energy is truly conserved.

# MASS-ENERGY EQUIVALENCE

The most famous of all physics equation did not arrive on the physics scene until Albert Einstein's theory of special relativity in 1905:

$$E = mc^2$$

energy = (mass)(speed of light)$^2$

Often, this equation is cited in the context of nuclear reactions or the weights of nuclei as an explanation of where the energy comes from during nuclear transformations (chemists talk about mass deficits). This equation, however, is much broader and more inclusive. It is stating that what we think of as mass (representing the inertia of a system) is simply the total energy contained within a system (kinetic and potential). Consider, for example, the mass of your coffee and its mug. Classically, the mass is considered constant as the temperature changes. Practically, there may be no appreciable change in its mass. However, as your coffee and mug cool down, they are reducing their mass as well—just not very significantly. Because the speed of light is such a large number, these changes in mass and energy can be ignored unless you are dealing with the high-energy, low-mass world of nuclear reactions.

The real question about mass that physicists have been worried about ever since the acceptance of Einstein's equation has been why some fundamental particles, which have no internal structure and thus no internal energy, have any mass at all. The mass of all macroscopic objects is due almost entirely to the energy within. (In fact, if you add up the masses of all the fundamental particles within a macroscopic object, that total mass is less than 1% of the total mass of the macroscopic object itself.) This residual mass of fundamental particles was resolved in 2012 with the discovery of the Higgs boson. The Higgs mechanism for giving fundamental particles mass relies on an interaction with a new type of field called the Higgs field, so even here we do not have mass in isolation.

With this proper lens for $E = mc^2$, the so-called mass deficit and energy released/absorbed from nuclear reactions is explained simply by the lesser and greater nuclear potential energies involved in the binding together of nucleons within the nucleus. Of course, this happens all the time, even for the lesser energy scale of exothermic and endothermic chemical reactions. However, the differences in mass are negligible for these lower-energy, higher-mass reactions.

# BONDING ENERGIES

Bond energy (gravitational bonds, electric bonds, and nuclear bonds) is simply negative potential energy. Often, these energies are represented simply as energy differences or are rescaled to appear positive. However, a great unifying principle is missed when not representing these energies as negative, as shown in Figure 6.1.

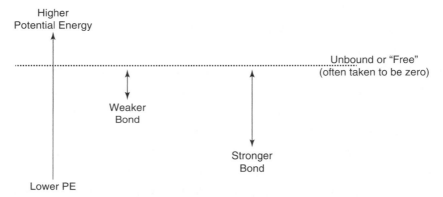

FIGURE 6.1 Bond energies.

Strong bonds are those that are hard to break, meaning a large negative potential energy is associated with them. Weak bonds, on the other hand, are easy to break: a small negative potential energy is associated with them. Consider the clarifying power of this view of binding energy with the following three energy-releasing examples:

1. A ball rolls downhill on Earth (gravitational bonds).

2. An adenosine triphosphate (ATP) molecule converts into adenosine diphosphate (ADP) in one of your cells (chemical bonds).

3. A U-235 nucleus undergoes fission (nuclear bonds).

In each case, a low negative potential energy (weaker bond) is being shifted to a deeper, more negative potential energy (stronger bond). Since energy is conserved, there must be a corresponding increase in other positive energies:

1. The ball-Earth system is moving from a weak bond (the ball and Earth are far apart) to a stronger bond (the ball and Earth are closer together).

2. The weaker bond between the third phosphate group and the rest of the molecule has been replaced with a stronger bond (between phosphate and water, for example).

3. The weaker bonds between the nucleons in the U-235 nucleus have been replaced by strong nuclear bonds in the tighter nuclei of the daughter atoms.

Table 6.3 summarizes the conservation laws.

### TABLE 6.3 CONSERVATION LAWS SUMMARY

| Linear Momentum | Angular Momentum | Energy |
|:---:|:---:|:---:|
| $\vec{p} = m\vec{v}$ | $\vec{L} = I\vec{\omega}$ | $ME = KE + PE$ |

If you have an isolated system, that is, one in which no outside forces are present, these quantities are conserved:

| | | |
|:---:|:---:|:---:|
| $\Sigma\vec{p_i} = \Sigma\vec{p_f}$ <br> or <br> $\Sigma m\vec{v}$ = constant | $\Sigma\vec{L_i} = \Sigma\vec{L_f}$ <br> or <br> $\Sigma I\vec{\omega}$ = constant | $\Sigma E_{initial} = \Sigma E_{final}$ <br> or <br> $\Sigma KE + \Sigma PE$ = constant |
| An object in motion stays in motion. | A rotating object continues to rotate. | Energy can be transformed but neither destroyed nor created. |
| The laws of physics are the same in any location. | The laws of physics are the same in any direction. | The laws of physics are the same at any time. |

If there is an outside force, these quantities are not conserved but are changed in a well-defined way:

| Impulse changes linear momentum. | Torques change angular momentum. | Work changes energy. |
|:---:|:---:|:---:|
| $\vec{F}\Delta t = \Delta\vec{p}$ | $\vec{\tau} = \dfrac{\Delta\vec{L}}{\Delta t}$ | $W = \Delta E$ |

# Chapter Review Exercises

1. Two cars collide.

    a. In what system is momentum conserved?

    b. In what system would you have to calculate a change in momentum due to impulse?

2. A sliding object is brought to a stop by a compressing spring.

    a. In what system is mechanical energy conserved?

    b. In what system would the work done by the spring be changing the kinetic energy of the object?

3. A cart smashes into a large wall of unknown mass. Should your system be the cart alone or the cart and the wall? Why?

4. The moon changes velocity as its elliptical orbit is being tracked.

    a. What is the system if the increase in speed as the moon heads toward perigee is being modeled as an increase in kinetic energy due to the work done by Earth's gravity?

    b. What is the system if this same increase is being modeled by a decrease in potential energy?

5. Mercury's rotational rate is slowly decreasing over time due to tidal locking with the sun.

    a. What is the system in which angular momentum is conserved?

    b. What is the system that explains this decrease in angular momentum as some kind of external torque being applied over time?

6.  Imagine a two-dimensional baseball field. The pitcher throws a baseball (mass = 0.145 kg) at a speed of 8.000 m/s straight at the batter. The batter provides a force for a brief period of time with his bat on the ball. As you know, this force multiplied by the time of contact between the ball and bat is called the impulse. The ball flies straight toward the shortstop at 10.00 m/s as shown below. The shortstop catches the ball.

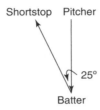

Shortstop    Pitcher

25°

Batter

a.  Write down $v_i$ and $v_f$ in component notation.

b.  Determine the components of the impulse the batter supplies.

c.  If the ball is in contact with the bat for 0.2000 seconds, what is the magnitude of force the batter applied?

7.  A high school pitcher is employing a ballistic pendulum to determine the speed of his fastball. A 5.6 kg lump of clay is suspended from a long cord. When the pitcher throws his fastball aimed at the clay, the ball becomes embedded in the clay and the two swing up to a maximum height of 0.15 m above the collision point. If the mass of the baseball is 0.15 kg, find the speed of the pitch. (Hint: First, calculate conservation of energy for the combined mass swinging. Then calculate conservation of momentum for the collision part. Mechanical energy is not conserved through the embedding of the ball in the clay; there are a lot of frictional forces at play during that time!)

# ELECTRICITY

## WHAT YOU WILL LEARN

- The nature of charge and how it is transferred.

- Charge distribution in conductors and insulators.

- How to calculate electric force with Coulomb's law.

- The relationship of electric energy and voltage.

| LESSONS IN CHAPTER 7 | |
| --- | --- |
| • Charge | • Coulomb's Law |
| • Conductors and Insulators | • Voltage |
| • Charge Transfer | |

## 7.1 Charge

Just as mass is a property of an object and indicates how the object will interact with gravitational forces, charge is also a property of matter and governs the object's interaction with electric and magnetic forces. Unlike mass, charge comes in two types: positive and negative. The positive charges in everyday matter can be traced back to the proton. The negative charges in common forms of matter can be traced back to the electron. These two subatomic particles carry equal and opposite amounts of charge. Since no smaller amounts of charge can be found in isolation in nature, this magnitude of charge is known as the elementary charge ($q_e$). Charge is measured in coulombs

(C), and every macroscopic amount of charge is made up of an integer multiple of the magnitude of this elementary charge. In other words, charge is quantized in the following fundamental units:

$$q_e = 1.6 \times 10^{-19}\text{ C}$$

Although some particles truly have no charge (e.g., photons and neutrinos), most matter is neutral because it contains equal amounts of both electrons and protons (so that the net charge is zero despite being full of charges). Because the protons are tightly bound by the strong nuclear force deep in the nucleus of their individual atoms, almost all charge movement in solids is accomplished by the movement of the electrons. Hence, although it may be convenient to say an object has acquired a positive charge, it most likely became positive by losing electrons rather than gaining protons. Since the nature of subatomic particles was unknown in the early days of electricity, early scientists unfortunately began the custom (which is continued to this day) of tracking the movement of positive charges. Even though the actual motion in circuits is negative charges heading in the opposite direction, physicists consistently defined their electric and magnetic relationships in terms of the movement of positive charge.

As with the more abstract quantities discussed in Chapter 6, total charge in an isolated system is a conserved quantity. Charge can be transferred, created, and destroyed but only such that the total amount of charge is the same before and after. (The symmetry underlying this conservation law is the complicated idea of gauge invariance for electromagnetism.)

EXAMPLE  After scuffing your feet on the carpet, you have accumulated a new charge of +8.6 μC (microcoulombs), which may cause you to experience a shock when you reach for the door handle. Did you gain or lose electrons while scuffing your feet on the carpet? How many?

**SOLUTION**

Since your net charge is positive, you lost electrons. To find out how many, simply divide your net charge by the fundamental charge on an electron:

$$\frac{8.6\ \mu\text{C}}{1.6 \times 10^{-19}\ \text{C/electron}} = \frac{8.6 \times 10^{-6}\ \text{C}}{1.6 \times 10^{-19}\ \text{C/electron}} = 5.4 \times 10^{13}\text{ electrons}$$

# 7.2 Conductors and Insulators

Just as every atom has a certain amount of affinity for its electrons (measured by the amount of energy required to free the electron from its atom), different materials have differing relationships to some of their valence electrons. Sometimes these electrons remain associated with each individual atom within the material. In this case, the material is considered to be a good insulator: it does not allow electrons to travel easily through it. Good insulators have high electric resistivity. Other materials tend to have some communal sharing of electrons, which allows its electrons to move relatively freely throughout the material. These materials are good conductors. Good conductors have low resistivity. This resistivity property of the material, along with the geometry of the item, determines its electric resistance (see Chapter 8 for details on resistivity and resistance).

If an object has an excess of one kind of charge, those charges will repel each other. In a conductor, these excess charges are free to move and thus will quickly move apart. As a general rule, charges in a conductor spread across the conductor's surface in such a way as to minimize their electric energy. A uniform conducting object, therefore, will have a uniform distribution of surface charge whenever it has excess electrons (or is lacking electrons). Conductors with edges will have a concentration of excess charge near the edges. See Figure 7.1.

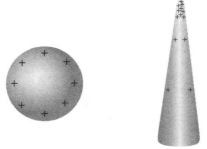

**FIGURE 7.1** Charge distribution in conductors.

## THINGS TO THINK ABOUT

If the object is positive and the protons are not free to move, how does the positive charge distribute itself uniformly over the surface? Although the protons themselves do not move, the electrons, which are free to move, will quickly redistribute themselves such that the "missing" spots are spread out as much as possible. In fact, the tracking of these holes in the electron coverage, which normally cancels out the positive charge of the nuclei, is, in essence, the movement of positive charges.

# 7.3 Charge Transfer

Charging by friction is accomplished when two neutral objects are rubbed against each other. If the two materials have a different affinity for their surface electrons, the one that has a lower affinity for its surface electrons will lose some electrons (thereby becoming positive) due to the frictional forces, while the other substance will gain those same electrons. Afterward, you will have two equally (but oppositely) charged objects.

For example, rub a balloon on your head. You will see that your hair and the balloon are now oppositely charged, as shown in Figure 7.2.

**FIGURE 7.2** Charging by friction.

Charging by conduction happens when a charged object is placed into contact with another object. The net charge is then shared over both objects. If the objects are good conductors, this will happen quickly and the excess charges will spread out on the combined surface of the objects. The two objects in contact will now carry the same types of charge, but the distribution of charge will depend on their relative sizes and resistivity. See Figure 7.3.

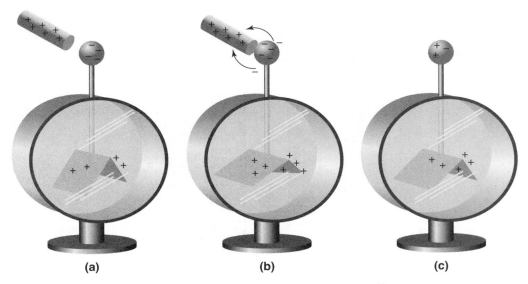

**FIGURE 7.3** Charging an electroscope by conduction.

Pictured here is a positively charged rod brought close to an electroscope (a), put into contact with the electroscope (b), and subsequently removed (c). An electroscope is simply an object designed to show charge: Since the two "leaves" of the electroscope are connected but somewhat free to move, any charge will be detected: (a) the nearby presence of charge causes the neutral electroscope to experience induced charge separation so that the leaves will separate slightly, (b) contact enables some charge transfer such that the leaves will be even more charged and thus separate more strongly, (c) without the presence of the rod, the charges will redistribute and thus the leaves will remain charged but less so than in (b).

If a charged object is brought into contact with a much, much larger object, after the charge is "shared" over the combined object, almost all the original excess charge will wind up on the larger object. This is known as **grounding** an object. An electric ground can be thought of as a bank for electrons rather than dollars. It will take any extra and supply you with them if you are missing some. The bank remains relatively neutral despite these transactions because the bank is so big. The largest object around in our daily lives is the planet Earth, hence the best "ground" is literally the ground itself!

Charging by induction involves several steps. The first is an important factor in understanding the static cling phenomenon. When a charged object is brought near a neutral object but not in contact with the neutral object, the charged object will affect the charges within the neutral object even though there is no charge transfer happening. Within the neutral object, the opposite charges will be attracted to the nearby charged object while the similar charges will be repelled. This induced charge separation leads to an overall attraction between the neutral object and the nearby charged object (for example, static cling). See Figure 7.4.

**FIGURE 7.4** The charged balloon causes the wood molecules to become polarized.

If the neutral object is grounded while induced charge separation is happening, the "ground" can easily take any electrons that are being repelled or supply any electrons that are being attracted. Because of this contact with the ground, the formerly neutral object will now carry the opposite charge of the nearby charged object that induced the charge separation in the first place. See Figure 7.5.

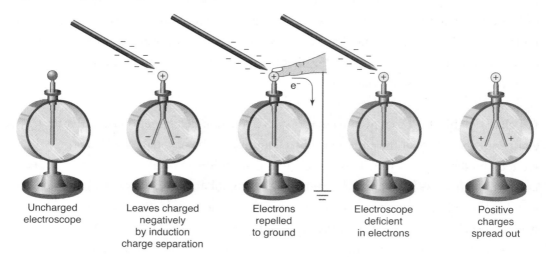

| Uncharged electroscope | Leaves charged negatively by induction charge separation | Electrons repelled to ground | Electroscope deficient in electrons | Positive charges spread out |

**FIGURE 7.5** The steps in placing a residual charge on an electroscope by induction.

# 7.4 Coulomb's Law

The actual force of electric attraction or repulsion between any two charges can be determined via **Coulomb's law**. Similar to Newton's universal gravity, it is an inverse square law and shares the same mathematical structure:

$$F = \frac{kq_1q_2}{r^2}$$

In this equation, $q_1$ and $q_2$ are the amounts of charge in coulombs and $r$ is the distance between those charges (from center to center) in meters.

The electric force constant $k$ has the value $8.99 \times 10^9$ N • m²/C² ($\approx 9 \times 10^9$ N • m²/C²). If we have a distribution of point charges, the net force on one charge is the vector sum (superposition) of each of the individual electrostatic forces a particular charge has with each other charge one at a time.

**EXAMPLE** Three charges are arranged in a row, as shown below. Find the net force on the leftmost charge.

**SOLUTION**

First, determine the direction of each force. The middle charge is negative, so it will repel the leftmost charge toward the left. The charge on the right is positive, so it will attract the leftmost charge toward the right.

Next, determine the magnitude of each force. Since we have already used the signs of the charges to determine direction, use the magnitude of charge only (absolute value) in Coulomb's law. Let $F_{12}$ mean the magnitude of force between the charge on the left and the charge in the middle. Let $F_{13}$ mean the magnitude of the force between the charge on the left and the charge on the right.

$$F_{12} = \frac{k\left(315 \times 10^{-6}\right)\left(215 \times 10^{-6}\right)}{(0.725)^2} = 1{,}160 \text{ N}$$

$$F_{13} = \frac{k\left(315 \times 10^{-6}\right)\left(225 \times 10^{-6}\right)}{(0.725 + 0.822)^2} = 266 \text{ N}$$

Finally, take the vector sum of the individual forces.

$$F_{\text{net}} = 266 - 1{,}160 = -894 \text{ N (to the left)}$$

## 7.5 Voltage

Voltage is an indirect measure of electric potential energy. Specifically, voltage is the electric potential energy available per charge:

$$\text{volt} = \text{joule/coulomb}$$

$$(V = J/C)$$

or

$$\Delta U = Q \Delta V$$

In other words, the change in electric potential energy ($\Delta U$) is the amount of charge ($Q$) transferred across a potential difference ($\Delta V$).

The difference in voltage between two locations (called potential difference, electric potential, or voltage drop) is what drives the charges to move from one location to another. To lower the electric potential energy of the system (and hence raise other forms of energy via conservation of energy), positive charges naturally move from higher voltage to lower voltage ($+Q$ with $-\Delta V$). Negative charges naturally move from lower voltage to higher voltage ($-Q$ with $+\Delta V$).

This is analogous to thinking about why objects naturally fall downward: they are lowering their gravitational potential energy.

**EXAMPLE**  Electric power has the same units as mechanical power (watts). The flow of charges is measured in amperes, which is the metric unit of current ($I$). If $P = IV$, then determine the units of ampere in base SI units.

**SOLUTION**

$$P = IV$$

$$\text{watts} = \text{amperes} \bullet \text{volts}$$

Substitute in our definitions of watts and volts:

$$\frac{\text{joule}}{\text{second}} = \text{ampere} \bullet \left( \frac{\text{joule}}{\text{coulomb}} \right)$$

Solve for ampere:

$$\text{ampere} = \text{coulomb/second}$$

$$(A = C/s)$$

# Chapter Review Exercises

1.  Determine the net force (magnitude and direction) on the upper-left charge pictured below (distances are center to center).

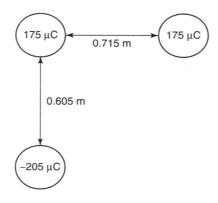

2.  How much energy does each electron that passes through the circuit take out of a 9-volt battery when that battery is being used to power a device?

3.  Explain in terms of charge and charge transfer why a dry sock that has just come out of the dryer might stick to your shirt.

4.  Explain how a lightning rod attached to your chimney might save your house from damage during a thunderstorm.

5.  A common household outlet in Europe has a 220-volt potential difference. Is energy being used even when nothing is plugged in? (The 220 volts is always present.)

# 8

# SIMPLE CIRCUITS

## WHAT YOU WILL LEARN

- What a simple circuit is and how charges move through it.

- The difference between resistance and resistivity.

- Ohm's law and how to calculate current.

- How to break down the current flow and energy distribution in series and in parallel circuits.

- How capacitors work.

| LESSONS IN CHAPTER 8 | |
|---|---|
| • Simple Circuits | • Kirchhoff's Laws for Circuits |
| • Resistance, Current, and Ohm's Law | • Circuit Diagrams |
| • Series and Parallel Circuits | • Capacitors |

## 8.1  Simple Circuits

An electric circuit is simply a conductive pathway (or several pathways) that allows the current to loop around (a complete path) so that the existing charges in the circuit itself are shuttling around the electric energy. At its simplest, a circuit consists of a source of electric power, a device that uses the electric power, and a conductive means of connecting them  (see Figure 8.1). The source of electric power is most commonly a battery for DC (direct current) circuits or is an outlet for AC (alternating current) circuits. (Note that an AC outlet ultimately draws its power from the power station powering the building.) The device that uses the electric power is most commonly a resistor.

**FIGURE 8.1** A simple circuit consisting of a battery, a resistor, three wires, and a switch.

 **THINGS TO THINK ABOUT**

*Where do the charges come from?*

Unfortunately, the language around circuits is littered with misinformation. For example, everyone says they are charging a battery or that a battery is losing its charge. However, the battery always has the same amount of charge: one electron per proton! In normal situations in an electric circuit, for every charge that enters a device, another charge is leaving that same device at the same time. What changes is the energy! Batteries are gaining or losing energy, not charge.

## 8.2 Resistance, Current, and Ohm's Law

An **insulator** requires a high potential difference in order to move charges through it, whereas a **conductor** requires very little. This variance in charge movement based on the material is known as the **resistivity** of a material. The actual resistance experienced in a certain physical situation is a function of the geometry of the pathway through the material that charges can take and the resistivity of the material:

$$R = \frac{\rho L}{A}$$

$R$ is the resistance of the object. It is measured in ohms, abbreviated by the Greek letter omega ($\Omega$).

The Greek letter rho ($\rho$) is the resistivity of the material the object is made from. It is measured in ohm-meters ($\Omega \cdot$ m).

$A$ is the cross-sectional area (measured in square meters, m$^2$) of the path the charges take through the object.

$L$ is the length (measure in meters, m) of the path through the object.

**EXAMPLE**   Compare the resistance of a cube of rubber with sides that are each 2 cm with that of a similarly sized piece of copper.

**SOLUTION**

The resistivity ($\rho$) of rubber is about $10^{12}$ $\Omega \cdot$ m, whereas that of copper is about $2 \times 10^{-8}$ $\Omega \cdot$ m.

Use the formula after converting centimeters into meters:

$$R_{rubber} = \frac{\left(10^{12} \ \Omega \cdot m\right)(0.02 \ m)}{(0.02 \ m)(0.02 \ m)} = 5 \times 10^{13} \ \Omega$$

$$R_{copper} = \frac{\left(2 \times 10^{-8} \ \Omega \cdot m\right)(0.02 \ m)}{(0.02 \ m)(0.02 \ m)} = 1 \times 10^{-6} \ \Omega$$

That difference of 20 orders of magnitude explains why, under everyday low-voltage situations, rubber effectively does not allow charge to flow through it (effectively infinite resistance), whereas copper does (effectively zero resistance).

If there is a source of easy-to-move charges (such as the conduction layer electrons in a metal), charge will flow if there is a potential difference. This flow of charge is known as **current** ($I$):

$$I = \frac{Q}{\Delta t}$$

In the equation above,

$I$ = current in amperes (amps);

$Q$ = amount of charge flowing in coulombs;

$\Delta t$ = time for the amount of charge ($Q$) to pass (in seconds).

If there is not a complete loop or any accumulation of charge, the flow of charge will not be continuous and will quickly come to an end. However, if a continuous loop is available, the charge will continue to flow as long as the potential difference is maintained. This is known as a complete circuit. The two most important complete circuits are ones with AC and those with DC. In an alternating circuit, such as one driven by a generator, the voltage is being switched at a steady frequency; the high and low voltages are swapping. That is why AC power has a frequency associated with it. (In the United States, 60 hertz is the frequency that power companies use.) The electric current then dances back and forth in the wires in response to this alternation. (Note that common wall outlets are themselves driven by generators and are thus outlets supply alternating voltage.) In a direct circuit, such as those driven by a battery, there is a fixed high-to-low voltage direction so the current never changes direction. Introductory circuits always use DC in their examples as it is easier to talk about; however, the basic principles are the same in either type of circuit.

## THINGS TO THINK ABOUT

*Why do we use AC?*

There are many reasons why we use AC for the distribution of electric power. However, one of the central reasons is that AC power is easier to transform to different voltages. Transformers are devices described in Chapter 10. They require constantly *changing* current in order to work. By transforming the voltages to be much higher for long-distance transmission, the same power transmission can be performed at much lower currents, which, in turn, means thinner wires and less loss of power during transmission.

The relationship between the resistance of the path followed by electric current and the potential difference driving that current is known as **Ohm's law**. (Objects that can be described by a fixed resistance and thus obey Ohm's law are known as ohmic.) Ohm's law is shown with the following equation:

$$\Delta V = IR$$

The electric energy "spent" is actually *transformed* into another form of energy. That form is usually thermal. The energy is no longer available to the electric circuit. The rate at which the electric energy is spent, called power, is determined by the potential difference (energy/charge) and the current (charge/second):

$$P = I\Delta V$$

In this formula, $P$ is the power in watts (joules/second), which is the same rate of energy transfer introduced in Chapter 6.

## THINGS TO THINK ABOUT

*Nonohmic devices*

Note that not all objects are ohmic; their relationship between potential difference applied and current flowing is not linear. Most important among these are the semiconductors, which are the backbone of digital technology. Briefly, semiconductors behave as high-resistance objects at lower voltages (off) and as good conductors (on) at higher voltages. This is the basis of the binary nature of digital technology.

**Switches** are used like drawbridges to control whether the circuit is complete or not. When a switch is closed, the circuit is complete and allows charges to flow. When a switch is open, the circuit is not complete and no charges can flow. Since each charge along the path is being pushed along by the previous charge on the path, once a switch is open anywhere in the loop, the current stops everywhere along the path. A **circuit breaker** is an automatic switch that is designed to open for safety reasons if the current passing through it becomes too high. A simpler form of this type of safety switch is a **fuse**. A fuse is a device that opens the circuit by actually burning out and must be physically replaced before the circuit is complete again.

A **short circuit** is a potentially dangerous situation where a complete pathway without a resistor is present. In this case, the usually negligible resistance of the circuit materials (wires, batteries, switches, etc.) themselves draw an enormous amount of current. This energy must be distributed somewhere, somehow, along the circuit. In a short circuit, items may melt or generate sparks, fires might start, and batteries can explode, hence the need for fuses and circuit breakers.

## THINGS TO THINK ABOUT

*How do the charges "carry" around the energy?*

A common misconception is that the energy of the charges is in the motion of the charges. Although charges are being displaced and are circulating in a complete circuit, the energy is carried in the fields associated with these charges (see Chapter 9), namely in the electric and magnetic fields between the charges (and, indeed, outside of the wires).

# 8.3 Series and Parallel Circuits

Circuit elements can either be on the same path or provide alternate paths between two points. If circuit elements are on the same path, they are said to be in **series** (see Figure 8.2). If they are on different paths, they are in **parallel** (see Figure 8.3). In a DC circuit, the same charges will travel, in turn, through each element in series. However, in a parallel circuit, each charge will take only one of the paths available. Two pathways in parallel are, in effect, connecting across the same potential difference, whereas items in series each have their own independent potential differences.

## THE EPE AS GPE MODEL OF CIRCUITS

When using the EPE (Electrical Potential Energy) as GPE (Gravitational Potential Energy) model of circuits, remember that the pipes are always full of water!

- Batteries are like pumps that raise the water (charges) to higher elevation, gaining energy.

- Resistors are where the water falls down to lower elevation, losing energy. Imagine the energy being harnessed by turning a paddle wheel rather than by any change in speed.

- Wires are like pipes carrying the water.

- Voltage is comparable to height.

- Current is like the flow of water.

Circuit resistance is similar to regulating the throughput of water.

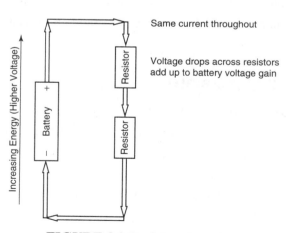

**FIGURE 8.2** Resistors in series.

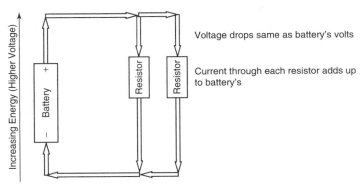

**FIGURE 8.3** Resistors in parallel.

# 8.4 Kirchhoff's Laws for Circuits

Two very important conservation laws underlie all circuit analysis: conservation of charge and conservation of energy. Gustav Kirchhoff first applied these to circuits and gave us the following rules for circuits.

1.  The junction rule: The total current coming into a junction must equal the current leaving the junction (charge must be conserved).

2.  The loop rule: The total voltage drops and gains must total to zero as you travel around any closed loop of a circuit (energy must be conserved). Traveling across a resistor with the current is a voltage drop; traveling against the current is a voltage gain. Traveling across a battery from negative to positive is a voltage gain; traveling across a battery from positive to negative is a voltage drop.

By applying Ohm's law along with Kirchhoff's laws, we can deduce how the resistances of the separate circuit elements combine to form a total resistance for the entire circuit. This total resistance is more properly referred to as the **equivalent resistance**. Resistors that are in series with each other simply accumulate more and more resistance to the current; they add. Resistors in parallel with each other, however, present more pathways for the current(s) to travel, so the equivalent resistance is actually lower—their conductances (inverse of resistance) sum! Circuits that are made of resistors entirely in parallel with each other or entirely in series with each other are known as simple circuits. Table 8.1 summarizes the rules of voltage, current, and resistance for simple circuits.

## TABLE 8.1 VOLTAGE, CURRENT, AND RESISTANCE IN SERIES AND PARALLEL CIRCUITS

| Series Circuits | Parallel Circuits |
|---|---|
| Voltage is additive. The sum of the voltage drops across the resistors in a series circuit is equal to the voltage supplied by the power source.<br><br>$$V_{source} = V_1 + V_2 + \ldots + V_f$$ | Voltage is constant. The voltage drop across each parallel component is equal to the voltage supplied by the power source.<br><br>$$V_{source} = V_1 = V_2 = \ldots = V_f$$ |
| Current is constant. The flow of current is the same in all parts of a series circuit.<br><br>$$I_{source} = I_1 = I_2 = \ldots = I_f$$ | Current is additive. The sum of the currents flowing through each parallel component equals the total current being supplied by the source.<br><br>$$I_{source} = I_1 + I_2 + \ldots + I_f$$ |
| Resistance is additive. The sum of the resistances in a series circuit equals the equivalent resistance.<br><br>$$R_{eq} = R_1 + R_2 + \ldots + R_f$$ | The reciprocal of the equivalent resistance is equal to the sum of the reciprocals of the resistances in parallel.<br><br>$$\frac{1}{R_{eq}} = \frac{1}{R_1} + \frac{1}{R_2} + \ldots + \frac{1}{R_f}$$ |

## 8.5 Circuit Diagrams

Simplified representations of a particular circuit can be drawn in several ways. Some of the most common symbols used are listed in Table 8.2. When drawing circuits, wires are assumed to have zero resistance, as are power sources. If the internal resistance of a wire or battery cannot be ignored, they can be represented by placing a small resistor in series with the battery or wire. Note that the actual physical placement of the components is not represented; only the relative series or parallel connections are represented. Real circuits are usually much messier to look at than their circuit diagram!

## TABLE 8.2  COMMON CIRCUIT SYMBOLS AND THEIR FUNCTIONS

| Circuit Element | Symbol | Function |
|---|---|---|
| DC power source | | Long side is the positive side (the higher-voltage side). Supplies the energy to create the current in the circuit. |
| AC power source | | Alternates high- and low-voltage sides to create an oscillating current. Supplies the energy to create the current in the circuit. |
| Resistor | | Transforms electric energy into thermal energy. |
| Switch | | Opens (stops current) or closes (completes the circuit). |
| Voltmeter | | Measures the voltage difference between two points. Ideally offers an infinite amount of resistance. |
| Ammeter | | Measures the current flowing through it. Ideally offers zero resistance. |
| Capacitor | | Temporary storage of charge and energy. |

Fill in the missing information in the chart for the following circuit:

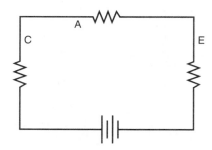

| Resistor | Resistance ($R$) | Current ($I$) | Voltage Drop ($V$) |
|---|---|---|---|
| $A$ | 25 Ω | | |
| $C$ | | 0.50 A | |
| $E$ | 75 Ω | | |
| Battery | N/A | | 60. V |
| Entire circuit equivalent | | N/A | N/A |

**SOLUTION**

First, identify the circuit as a series circuit: there is only one path for the current.

Since the current must be the same for all elements in series, we know that each entry in that column must be 0.50 A.

Knowing $I$ and $R$ for resistors A and E allows us to use Ohm's law to determine the voltage drop for each:

$$\text{Resistor A: } V = IR = (0.5)(25) = 12.5 \text{ V}$$

$$\text{Resistor E: } V = IR = (0.5)(75) = 37.5 \text{ V}$$

Using the loop rule, we know that there are 60 volts available for all three resistors to use. Since 12.5 V + 37.5 V = 50 V is being used by resistors A and E, we can determine the volts used by resistor C:

$$60 - 50 = 10 \text{ V}$$

Now we can used Ohm's law for resistor C:

$$V = IR$$

$$10 = 0.50R$$

$$R = 20 \text{ ohms}$$

We can now find the equivalent circuit resistance in two ways. First, add the resistances:

$$25 + 20 + 75 = 120 \text{ ohms}$$

Second, we can apply Ohm's law to the entire circuit:

$$V = IR$$

$$60 = 0.50R$$

$$R = 120 \text{ ohms}$$

They match!

| Resistor | Resistance ($R$) | Current ($I$) | Voltage Drop ($V$) |
|---|---|---|---|
| $A$ | 25 Ω | 0.50 A | 12.5 V |
| $C$ | 20 Ω | 0.50 A | 10 V |
| $E$ | 75 Ω | 0.50 A | 37.5 V |
| Battery | N/A | 0.50 A | 60. V |
| Entire circuit equivalent | 120 Ω | N/A | N/A |

Fill in the missing information in the chart for the following circuit:

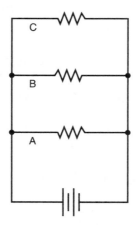

| Resistor | Resistance (R) | Current (I) | Voltage Drop (V) |
|---|---|---|---|
| A | 18 Ω | | |
| B | | 0.75A | |
| C | 4.5 Ω | | |
| Battery | N/A | | 4.5 V |
| Entire circuit equivalent | | N/A | N/A |

**SOLUTION**

First, identify this as three resistors in parallel, as each one lies upon a different path and will receive different currents.

Since the resistors are all in parallel, they all experience the same voltage drop: 4.5 V.

Applying Ohm's law for each resistor allows us to fill in the missing info for each:

$$\text{Resistor A: Current} = \frac{V}{R} = \frac{4.5}{18} = 0.25 \text{ A}$$

$$\text{Resistor B: Resistance} = \frac{V}{I} = \frac{4.5}{0.75} = 6 \text{ Ω}$$

$$\text{Resistor C: Current} = \frac{V}{R} = \frac{4.5}{4.5} = 1 \text{ A}$$

Using the junction rule, the battery must supply all paths with their currents:

$$0.25 + 0.75 + 1 = 2 \text{ A}$$

There are two ways to find the equivalent circuit resistance. First, we can use Ohm's law for the entire circuit:

$$V = IR$$

$$4.5 = 2R$$

$$R = 2.25 \text{ ohms}$$

Second, we can add the reciprocals of the resistances:

$$\frac{1}{R} = \frac{1}{18} + \frac{1}{6} + \frac{1}{4.5}$$

$$R = \frac{18}{8} = 2.25 \text{ ohms}$$

| Resistor | Resistance ($R$) | Current ($I$) | Voltage Drop ($V$) |
|---|---|---|---|
| $A$ | 18 Ω | 0.25 A | 4.5 V |
| $B$ | 6 Ω | 0.75A | 4.5 V |
| $C$ | 4.5 Ω | 1 A | 4.5 V |
| Battery | N/A | 2 A | 4.5 V |
| Entire circuit equivalent | 2.25 Ω | N/A | N/A |

Complex circuits are made up of multiple pathways and/or multiple batteries. To solve these problems, employ the following two techniques.

1. Simplify any smaller portions of the circuit that are made of strictly in-series or in-parallel resistors. Redraw a simplified version of the circuit with which to work.

2. Assign an arbitrary direction of current in each path in the circuit. Apply Kirchhoff's laws to each junction and to as many loops as needed until you can solve for all the unknowns. If you guessed the wrong direction of current in the beginning, you will discover a negative value for that current. This will inform you that the direction of actual current flow is in the opposite direction from the one originally assigned.

# 8.6 Capacitors

Unlike the circuit elements we have discussed thus far, a capacitor is a circuit element that actually does collect charge. Although commonly represented by two parallel plates, in actual practice most real capacitors are built from two concentric cylinders. As charge collects on one side of the capacitor, the other side accumulates the opposite charge, maintaining neutrality. In this manner, the circuit appears to be complete even though there is a gap inside of the capacitor. For example, as an electron arrives on one side of a capacitor, it will push an electron away on the opposite side, revealing a positive charge there. However, as charge collects, it becomes increasingly difficult to place additional charges on the already charged capacitor. This results in a decreasing current passing through the capacitor. The following sequence of pictures in Figure 8.4, 8.5, and 8.6 illustrates the changing behavior of a pathway in a circuit that contains a capacitor.

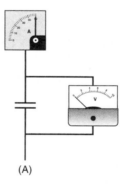

(A)

**FIGURE 8.4** Uncharged capacitor.

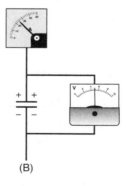

(B)

**FIGURE 8.5** Partially charged capacitor.

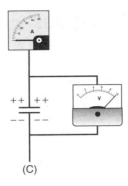

(C)

**FIGURE 8.6** Fully charged capacitor.

- Uncharged capacitor (Figure 8.4): The current appears to pass right through the separated plates as opposite charge just begins to accumulate. The pathway is effectively a closed switch at this point.

- Partially charged capacitor (Figure 8.5): As charges accumulate, a potential difference is generated across the gap that is opposing the overall voltage (not pictured) driving the current. The flow of current into and out of the capacitor diminishes as a result. (Note that the amount of charge remains the same on both sides. However, for each negative charge added to one plate of the capacitor, a positive charge is generated on the opposite plate.) The capacitor voltage is rising exponentially, so the current falls exponentially as a result.

- Fully charged capacitor (Figure 8.6): When the voltage across the capacitor is as large as the external driving voltage, the current ceases to flow and the pathway is effectively an open switch.

The capacitance of a capacitor is measured in farads (F), defined as the charge that can be stored per volt. The capacitance is determined by the manufacturing of the capacitor (what it is made of and the geometry of its plates) and can be determined using the following equation:

$$C = \frac{Q}{V}$$

The time dependence of the circuit is affected by both the resistance and the capacitance of the circuit. For this reason, these circuits are known as RC circuits. The time dependence of the charging, current, and voltage is exponential in nature. For example, the current falls off as the capacitor charges according to the following:

$$I_t = I_{max}e^{-\frac{t}{RC}}$$

1. Who gets power? In the table below, determine if R1 and/or R2 receive power from the battery in the switch configurations described in each row. Write "YES" if they get power and "NO" if they do not get power.

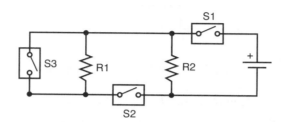

| S1 | S2 | S3 | R1 | R2 |
|--------|--------|--------|----|----|
| Open | Open | Open | | |
| Open | Closed | Closed | | |
| Closed | Open | Closed | | |
| Closed | Closed | Closed | | |
| Closed | Closed | Open | | |
| Open | Open | Closed | | |

2. Fill in the missing information in the table below about this circuit.

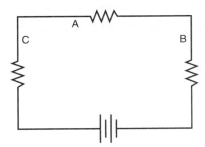

| Resistor | Resistance ($R$) | Current ($I$) | Voltage Drop ($V$) |
|---|---|---|---|
| $A$ | 25 Ω | | |
| $B$ | 50 Ω | | |
| $C$ | | | |
| Battery | N/A | | 300 V |
| Entire circuit equivalent | 150 Ω | N/A | N/A |

3. Fill in the missing information in the table below about this circuit.

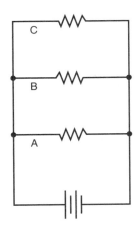

| Resistor | Resistance ($R$) | Current ($I$) | Voltage Drop ($V$) |
|---|---|---|---|
| $A$ | 18 Ω | | |
| $B$ | | | |
| $C$ | 27 Ω | | |
| Battery | N/A | 1.2 A | 9.0 V |
| Entire circuit equivalent | | N/A | N/A |

4.  The exponential rise and fall of current through and voltage across capacitors in series with resistors has an expression that contains the quantity RC (ohms • farads). Determine what SI unit this reduces to.

# FIELDS

## WHAT YOU WILL LEARN

- How forces are transmitted across space.

- How to draw and interpret gravitational and electric field lines.

- How to represent energy potentials in fields.

- How magnets and magnetic fields and forces work.

- How to apply various forms of the right-hand rule.

- How to measure the speed and types of charges traveling in wires.

| LESSONS IN CHAPTER 9 | |
| --- | --- |
| • Forces and Fields | • Magnetic Fields |
| • Gravitational Fields | • Magnetic Forces |
| • Electric Fields | • The Hall Effect |
| • Energy and Fields | |

## 9.1 Forces and Fields

Isaac Newton himself was bothered by something almost magical about some forces: How do they exert their influence on another object when they are not in contact? This conundrum of an "action at a distance" was so bothersome that physicists used to divide forces into contact forces (like friction and tension) and noncontact forces (like gravitational and electric forces). Nowadays, with our understanding of the fundamental forces and our knowledge of fields, we can answer the riddle of how all forces operate: the source of the force creates a field around it. A **field** can be thought of as a set of numerical values at all locations in space (like a contour map has elevation data at every point). The recipient of the force is sitting in this field and experiences a force because of it. As an example, Earth does not experience a gravitational pull from the sun directly; rather, the sun has created a gravitational field throughout space that has a certain value 93 million miles away, where Earth is located. It is the sun's gravitational field at this point (rather than the sun itself) that pulls on Earth.

Of the many fields that modern physics employs, the three classical fields are the gravitational, the electric, and the magnetic fields. With these three fields, all the forces can be understood, with the exception of the strong and weak nuclear forces (which have their own fields and interactions).

We have already examined gravitational and electric forces. So let's begin with these and see how the direct "action at a distance" equations can be recast as fields. In Table 9.1, we begin by describing the two forces in the traditional form. On the next line, we describe the fields being generated by the object considered as the source. (Note that which object is designated as the source and which is designated as the second object is arbitrary. Newton's third law still applies. The results will be the same if the roles are reversed.) The source object has mass $M$ and charge $Q$. Note that these field formulas can be thought of as tagging every point in space with a specific value, depending on that location's distance from the source ($r$). Technically, these are force fields because they have direction as well as magnitude at every location. The final line shows how the second object responds to the field when that second object is placed into the field. Note that the final equation is, of course, the same mathematically as the first; none of our previous work in other chapters is incorrect mathematically!

**TABLE 9.1 GRAVITATIONAL AND ELECTRIC FORCES**

|  | Gravitational Forces | Electric Forces |
|---|---|---|
| Force (N) between two objects | $F_g = -\dfrac{Gm_1m_2}{r}$<br><br>Newton's law of universal gravitation | $F_E = \dfrac{Kq_1q_2}{r^2}$<br><br>Coulomb's law |
| Force field from one object | $g = -\dfrac{GM}{r^2}$<br><br>Gravitational field in N/kg | $E = \dfrac{kQ}{r^2}$<br><br>Electric field in N/C |
| Force (N) on second object in field | $\vec{F} = m\vec{g}$ | $\vec{F} = q\vec{E}$ |

Sign convention: negative = attracting; positive = repelling

## THINGS TO THINK ABOUT

In case you think this idea of fields is too strange a model to be true or is unnecessarily complicated, know that the existence of fields is the underpinning of everything in modern physics. The particles themselves as we think of them classically have disappeared in quantum field theory—everything is now made of fields (see Chapter 15).

# 9.2 Gravitational Fields

How should one picture these fields? You should consider these fields as real rather than just as mathematical formalism! Figure 9.1 employs the idea of field lines (arrows) to indicate both the direction and the strength of the gravitational field being generated by Earth in the space around it. The arrowheads indicate the direction of the field; they show the direction of the gravitational force on a second object if that second object was placed into the field. The spacing between the lines indicates the relative strength of the field. The closer the field lines are together, the stronger the force is.

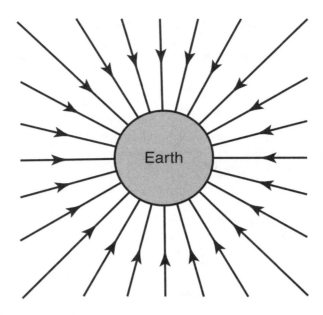

FIGURE 9.1 The direction and strength of the gravitational field generated by Earth.

## 9.3 Electric Fields

In the case of electric fields, our idea of field lines is complicated by the fact that the direction of the force is dependent on the type of charge (positive or negative) the second object possesses. As they have with everything regarding charge in electricity and magnetism, scientists have standardized their drawings. They draw the field lines acting on the second charge as if that second charge had a positive charge. A negative charge would simply have arrows pointing in the opposite direction. See Figure 9.2.

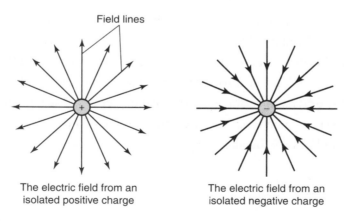

FIGURE 9.2 Electric field lines around positive and negative point charges.

Note the similarity of the negative charge's field to that of Earth's in Figure 9.1. A negative charge attracts a positive charge in much the same manner as a mass attracts a second mass. This is the reason that Coulomb's law and Newton's law of universal gravitation have such similar algebraic forms. Electric source charges are not usually isolated and can assume a variety of configurations. As a result, the electric field lines can be quite varied. The rules of thumb for sketching these field line configurations are as follows and can be seen in Figure 9.3.

- Field lines always begin on positive charges (sources) and end on negative charges (sinks).

- The number of field lines emerging from a source or going into a sink is proportional to the amount of charge at that location.

- Field lines never cross because there can be only one direction for the net electric force at any one location.

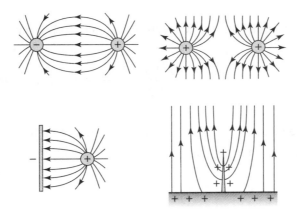

**FIGURE 9.3** Electric field lines.

Since all field lines would actually be altered by the presence of a second charge (or second mass in the case of gravitational fields), we generally use (or imagine using) a second charge that is much, much smaller than the existing charges for which we are mapping the field lines. This much smaller, second charge is known as the **test charge**. The existing charges creating the fields drawn are known as the **source charges**.

**EXAMPLE**  Determine the gravitational field at the surface of Earth using the following formula:

$$g = \frac{GM_E}{R_E{}^2}$$

**SOLUTION**

Plug in the values of the universal gravitational constant, the mass of Earth, and the radius of Earth:

$$g = \frac{\left(6.67 \times 10^{-11}\right)\left(5.97 \times 10^{24}\right)}{\left(6.37 \times 10^{6}\right)^2} = 9.81 \text{ N/kg}$$

Now that looks familiar!

**EXAMPLE**  If a single electron experiences an electric force of $2.2 \times 10^{-12}$ N to the east, what is the magnitude and direction of the electric field exerting this force?

**SOLUTION**

Using the charge on a single electron and the given force:

$$E = \frac{F}{q} = \frac{2.2 \times 10^{-12}}{1.6 \times 10^{-19}} = 1.4 \times 10^{7} \text{ N/C}$$

Since an electron is negative and it is experiencing a force to the east, the electric field is to the west.

# 9.4  Energy and Fields

Table 9.2 summarizes gravitational and electric energy fields.

**TABLE 9.2  A SUMMARY OF GRAVITATIONAL AND ELECTRIC ENERGY FIELDS**

| | Gravitational Fields | Electric Fields |
|---|---|---|
| Potential energy (in joules) between two objects | $U_g = -\dfrac{Gm_1m_2}{r}$<br><br>Gravitational potential energy | $U_E = -\dfrac{Kq_1q_2}{r}$<br><br>Electric potential energy |
| Energy field from one object | $\Phi = -\dfrac{GM}{r}$<br><br>Gravitational potential in J/kg | $V = \dfrac{kQ}{r}$<br><br>Electric potential (voltage) in J/C |
| Energy (in joules) when second object is in field | $U_g = m\Phi$ | $U_E = qV$ |

Sign convention: negative = attracting; positive = repelling

Energy fields are scalar fields. They are usually represented by drawing lines of equal value (**equipotential lines**). Equipotential lines intersect lines of force at right angles. Figure 9.4 shows an example of the electric force field and the voltage equipotential lines in a capacitor.

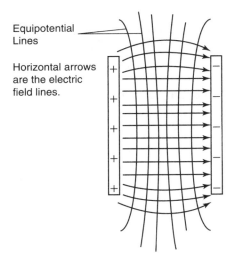

**FIGURE 9.4** Energy field represented by equipotential lines.

# 9.5 Magnetic Fields

In addition to electricity, another force is based on charges—magnetism. Just as charges create electric fields that other charges then respond to, *moving* electric charges create magnetic fields that other *moving* electric charges respond to. With the insights of Maxwell's equations about electromagnetism in the late 1800s and Einstein's insights about relativity in the early 1900s, our modern understanding is that magnetism and electricity are really the same basic charge-based phenomena seen from two different points of view. Nonetheless, magnetism presents itself as its own unique force from a reference frame in which charges are in motion. Unlike electricity and gravity, there really is no simple model for magnetism that does not involve an understanding of fields.

For centuries, people have used naturally occurring magnetic materials (lodestones) for navigation. All materials have magnetic properties, but nickel, cobalt, and iron are all naturally occurring, relatively strong magnets that can be found on Earth. The rules of interaction for these "permanent" magnets are deceptively similar to that of electric charges (see Figure 9.5).

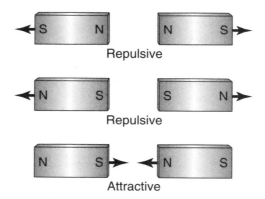

**FIGURE 9.5** Repulsion and attraction between magnets.

## THINGS TO THINK ABOUT

*How does a compass work?*

The entire Earth has a magnetic field due to magnetic material deep within. The compass is a permanent magnet. The compass's north pole is attracted to Earth's magnetic south pole, which is constantly moving as Earth's interior changes over time. However, for the past several thousand years, Earth's magnetic south pole has been near the actual rotational north pole (true north) of the planet. This begs the question: how did the north pole get its name? Originally, the north pole of a compass magnet was called the north-*seeking* pole!

The similarity of the two electric charges and the two sides of a magnet (north and south poles) quickly disappears upon investigation. In fact, the north and south poles of a magnet cannot be isolated and are not associated with different particles. If you break a magnet in half, you will be left with two smaller magnets, each with their own north and south poles (see Figure 9.6). You could continue this game all the way down to a single atom, which would still present its own north and south magnetic poles to the world.

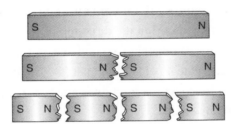

**FIGURE 9.6** Breaking a magnet produces smaller magnets, each with a north and a south pole.

The individual electrons within each atom are the charges in motion that are creating the magnetic field. The overall magnetic poles of a physical (permanent) magnet is the net effect of all the individual magnetic moments of all the electrons' magnetic fields being summed. How does a single charge in motion give rise to both a north and a south pole? This can be understood only in terms of the magnetic field generated by a charge in motion. A moving charge can be represented as a current as shown in Figure 9.7. The charge's motion creates a magnetic field line that loops around the motion of the charge.

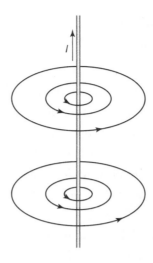

**FIGURE 9.7** A moving charge represented as a current.

As with the electric field, there is some ambiguity in defining the direction of the magnetic field lines. Namely, which way do the magnetic field lines loop: clockwise or counterclockwise? We have adopted a standard known as the right-hand rule. If the charge in motion in positive, point the thumb of your right hand in the direction of the charge's motion. Your fingers will curl in the direction of the magnetic field. As always, a negative source charge will do the opposite. To understand the north and south poles of a permanent magnet, we must create an electromagnet (a current-based magnet) by causing the source current to go in a loop. As the field lines attempt to circle around the current, the field shown in Figure 9.8 is created.

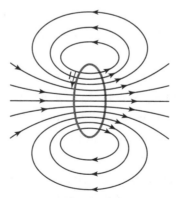

**FIGURE 9.8** Magnetic field lines found in an electromagnet.

Aha! Here is the answer to the dilemma of the north and south poles being inseparable! The north pole is the side where the magnetic field lines are coming out of the magnet, whereas the south pole is the opposite. In Figure 9.9, we are looking at the same field from a different point of view.

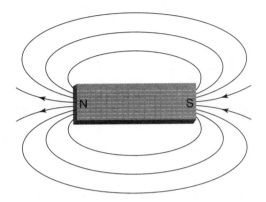

FIGURE 9.9 Magnetic field lines for a permanent magnet.

Note that there are no sources or sinks of magnetic field lines; they loop around. If put another way, there are no magnetic monopoles—for every north pole, there has to be a south pole too! The loop of magnetic field for a permanent magnet is completed within the magnetic material as indicated in Figure 9.9. A stronger electromagnet can be created in several ways: create many loops of current, increase the current, and fill the loop with magnetic material (see Figure 9.10). These devices are also referred to as solenoids. Not only can an electromagnet be easily adjusted in strength by changing the current, but it can be caused to flip its north and south poles simply by switching the direction of current.

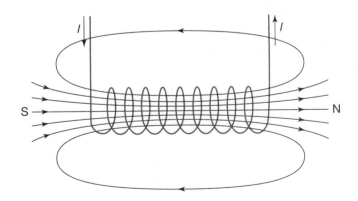

FIGURE 9.10 An electromagnet.

## THINGS TO THINK ABOUT

Determine the direction of the fields for an electromagnet using a different form of the right-hand rule, as shown in Figure 9.11. Grab the electromagnet with the fingers of your right hand curling in the direction of current flow; your thumb will point in the direction of the north pole (i.e., in the direction of the magnetic field).

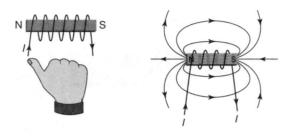

**FIGURE 9.11** Using the right-hand rule with an electromagnet.

## THINGS TO THINK ABOUT

If every electron in every atom is producing magnetic fields, why are some materials magnetic while some are not? Actually, all materials are magnetic to a certain extent. The question is, why are some materials strongly magnetic whereas others are less so? The answer is a combination of the exact configuration of electrons in the element (how many electrons have magnetic fields that are lined up within a single atom) and how the atoms within a material are magnetically aligned with their neighbors. The best permanent magnets are composed of magnetic atoms (containing multiple lined-up electron magnets) that are themselves magnetically aligned. By getting these magnetic atoms aligned or unaligned, you can create or destroy a permanent magnet.

The strength of a loop of magnetic field ($B$) around a straight wire carrying current ($I$) is calculated as follows:

$$B = \frac{\mu_0 I}{2\pi r}$$

In this formula, $\mu_0$ is the permeability of free space ($1.26 \times 10^{-6}$ m • kg • s$^{-2}$ • A$^{-2}$). It is also called the magnetic force constant and is playing a role similar to $G$ for gravitational forces or $k$ for electric forces. Units of magnetic fields are teslas (T).

**EXAMPLE** Rewrite teslas in base SI units.

**SOLUTION**

Extract the units from $B = \dfrac{\mu_0 I}{2\pi r}$:

$$\text{tesla} = \text{m kg s}^{-2}\,\text{A}^{-2}\,(\text{A})/\,\text{m} = \text{kg s}^{-2}\,\text{A}^{-1}$$

## THINGS TO THINK ABOUT

Note that the electric force constant, $k$, used in Coulomb's law is a simplified version of an expression involving a more fundamental electric constant, $\varepsilon_0$:

$$k = \dfrac{1}{4\pi\varepsilon_0}$$

In this equation, $\varepsilon_0$ is called the permittivity of free space. For both magnetism and electricity, the values of the permeability and permittivity of free space (as indicated by the subscript zero) are only for empty space. Fields in places other than a vacuum have different electric and magnetic constants.

# 9.6 Magnetic Forces

Magnetic forces can be calculated using two different formulas:

$$F = qvB\sin\theta$$

or

$$F = IlB\sin\theta$$

A moving charge ($qv$) or a length ($l$) of current ($I$) in a magnetic field ($B$) experiences a force ($F$) at right angles to both the field and the motion. These expressions for magnetic forces are examples of vector cross multiplication since $v$ and $B$ (or since $l$ and $B$) are both vectors. Two vectors make a plane, so the right-hand rule is used to determine in which direction away from the plane the force is. If the motion and the field are in the same direction ($\theta = 0$), there is no force.

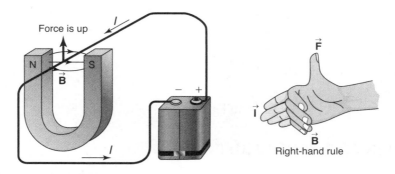

**FIGURE 9.12** Direction of magnetic forces.

Note that there is some ambiguity when drawing pictures in three dimensions, such as the pictures in Figure 9.12, when we are demonstrating the right-hand rules. To clarify directions, a shorthand representation for arrows coming into and out of the page has been codified through the years, as indicated in Figure 9.13. It shows a cross-sectional slice of an electromagnet (such as the one pictured in Figure 9.12) and the magnetic field produced within. In drawings of vectors in three dimensions:

- A dot is the tip of an arrow pointed at you ("out of paper").

- An X is the tail of an arrow pointed away from you ("into paper").

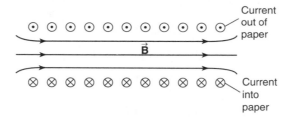

**FIGURE 9.13** The magnetic field produced by an electromagnet.

**EXAMPLE** Pictured below is the circular path that results from a charged object traveling in a perpendicular magnetic field. Derive an expression for the radius ($r$) of the circular motion of the electron in terms of the magnetic field ($B$), speed of electron ($v$), mass of electron ($m$), and charge ($q$).

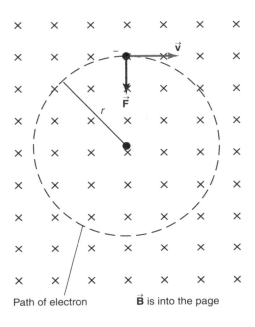

Path of electron      $\vec{B}$ is into the page

## SOLUTION

The force on the moving electron is directed inwardly as indicated by the right-hand rule. (Since the electron's charge, $q$, is negative, you must reverse the direction of the force from the right-hand rule or else use your left hand.) Calculate the magnitude of the magnetic force:

$$F = qvB$$

Since this force is causing the circular motion, it is a centripetal force:

$$qvB = F_c = ma = ma_c = \frac{mv^2}{r}$$

Solve for $r$:

$$r = \frac{mv}{qB}$$

Note that $m$ and $q$ are the mass and charge, respectively, of the moving charge.

**EXAMPLE** Determine the force of attraction per unit length for the current-carrying wires pictured below.

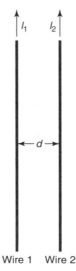

Wire 1    Wire 2

## SOLUTION

There will be a magnetic force because each current is creating its own magnetic field. One wire is sitting in the magnetic field of the other wire. By Newton's third law, the wires should be exerting equal and opposite forces— but that can also be shown by running through the following argument again, starting with the second wire and reversing all the roles.

To determine the force on wire 1, first we must determine the direction of the magnetic field from wire 2 at the location of wire 1. By the right-hand rule, we can quickly determine that the magnetic field lines produced by wire 2 loop around it such that they are coming *out of the page* in the region where wire 1 is located. The strength of the magnetic field is given by the following:

$$B = \frac{\mu_0 I}{2\pi d}$$

The field will exert a force to the right (toward wire 2) on wire 1 by the right-hand rule. Calculate the magnitude of that magnetic field:

$$F = I_1 l B = \frac{I_1 l \mu_0 I_2}{2\pi d}$$

Dividing by the arbitrary length of wire, $l$, we obtain the force per unit length:

$$\frac{F}{l} = \frac{\mu_0 I_1 I_2}{2\pi d}$$

# 9.7 The Hall Effect

Is a current made of positive charges moving in one direction or is it made of negative charges moving in the other direction? And how fast are these charges moving? So far, we have not explained how to determine the answers to these questions. Instead, we have relied on our knowledge of the structure of the atom to suppose that electrons are moving around rather than protons. With the addition of magnetic fields, we now have a definitive experiment that can be performed to prove what type of charges are moving in a current and at what speed. This experiment was first done by Edwin Hall in 1879 and thus bears his name to this day.

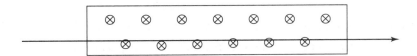

**FIGURE 9.14** A wire carrying current to the right.

Figure 9.14 shows an expanded view of a cross section of wire, or conductor, carrying current to the right. We have placed this section of wire in an external magnetic field (indicated by the X's in figure 9.14) that is passing through the wire into the page.

There are two possibilities for the current.

1.  Positive charges are moving to the right.

2.  Negative charges are moving to the left.

Since these charges are moving in a magnetic field, the right-hand rule can be used to determine the direction of the magnetic force. For either charge, the result is that the charge will be forced to the upper surface of the wire. Now imagine using a voltmeter to measure the potential difference between the top and bottom surfaces. There are two possible results.

1.  If the current is made of positive charges, the upper surface will be at higher potential than the lower.

2.  If the current is made of negative charges, the upper surface will be at a lower potential than the lower.

In fact, this is the definitive test that proves most currents are made of negative charges traveling in the opposite direction to the "current."

If we allow the experiment to continue, more charge will build up, creating an electric field pointed upward within the wire. When the voltage stops rising, the electric force downward on the electrons must be exactly equal to the magnetic force upward, preventing any further charge accumulation. Balancing these forces gives us the following:

$$qE = qvB$$

The charge ($q$) is the same on both sides, the electric field ($E$) is being measured indirectly with the voltmeter, and the magnetic field ($B$) is determined by the experiment externally and so is also known. This allows you to solve for the only unknown left, the velocity of the charge carriers in the wire. This is the experimental method for determining the "drift" velocity of charge carriers in a conductor:

$$v = \frac{E}{B}$$

The measured drift velocities of electrons in conductors carrying normal currents is surprisingly small: less than a mm/s.

# Chapter Review Exercises

1. An alpha particle (2 neutrons + 2 protons) enters a uniform magnetic field as indicated below. The initial velocity of the particle is 2,000 m/s, and the uniform magnetic field is 5 teslas downward.

$$
\begin{array}{ccc}
\times & \times & \times \\
\times & \times & \times \\
\times & \times & \times \\
\times & \times & \times \\
\end{array}
$$

$\xrightarrow{v}$

a. Sketch the path of the particle above, labeling $v$ (instantaneous velocity) and $F$ (force) at three points.

b. Find the radius of the resulting path.

c. If the particle was an electron instead of an alpha particle, how would your answer to part (a) change?

2. Two parallel wires are carrying the following currents in opposite directions: $I_1 = 2$ amps and $I_2 = 5$ amps. The wires are exactly 1 meter apart.

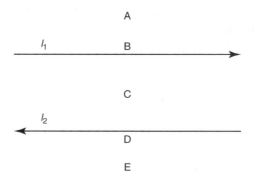

a. What is the net magnetic field (magnitude and direction) at each of the points labeled $A$, $B$, $C$, $D$, and $E$ above? Point $C$ is exactly halfway between the wires. Points $B$ and $D$ are on the wires themselves. Point $A$ is 0.5 meters above the top wire, and point $E$ is 0.5 meters below the bottom wire.

b. What is the force (magnitude and direction) per unit length at points $B$ and $D$ (in N/m)?

3. Points $A$, $B$, $C$, and $D$ are four points in an electric field.

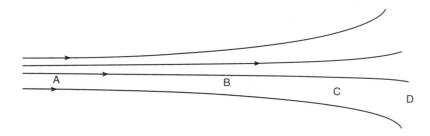

a. At which point is the electric field the strongest?

b. If you placed a proton at point $B$, which other point ($A$, $C$, or $D$) would it first encounter?

c. If you placed an electron at point $B$, which other point ($A$, $C$, or $D$) would it first encounter?

d. Which point has the highest voltage?

e. Which point has the lowest voltage?

f. Somewhere to the left of the picture there must be what kind of charge to create these field lines?

g. Draw two equipotential lines, one between points $A$ and $B$ and the other between points $C$ and $D$. Make sure both equipotential lines extend to the edge of the picture.

h. Why do all four points in space have voltage but no electric potential energy?

4. If a proton experiences an electric force of $3.2 \times 10^{-18}$ N, what is the electric field strength there?

# ELECTROMAGNETISM

## WHAT YOU WILL LEARN

- What flux is and how to calculate it.

- How speakers and electric motors work.

- How electric and magnetic fields are related to each other.

- How generators and transformers work.

- What electromagnetic radiation is, how it is generated, and the spectrum of possibilities.

- The relationship between Einstein's special relativity and electromagnetism.

| LESSONS IN CHAPTER 10 | |
|---|---|
| • Faraday and Lenz | • Electromagnetic Spectrum |
| • Common Electromagnetic Devices | • Special Relativity and Electromagnetism |
| • Maxwell and Electromagnetic Radiation | |

## 10.1  Faraday and Lenz

In 1831, Michael Faraday made a series of observations in his lab in which an electric field was created by manipulating magnetic fields. This relationship, which underpins the entire electric world we now live in, is called Faraday's law of **induction**. The essential phenomenon is that *changes* in magnetic flux induce an electric field. The **flux** of a field ($\Phi_B$) is defined as the strength of the field crossing a defined area (see Figure 10.1).

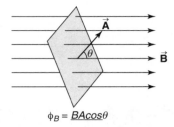

$$\phi_B = BA\cos\theta$$

**FIGURE 10.1** Faraday's law.

This induced potential difference is sometimes called EMF for electromotive force and is a synonym for voltage. The induced potential difference driving the electric field is equal to the rate of change of the magnetic flux:

$$V_{induced} = \text{EMF} = \frac{\Delta\Phi_B}{\Delta t}$$

The presence of a magnetic field is not enough to create the potential difference (or electric field); the magnetic field must be changing. More specifically, the magnetic field must be changing in strength or orientation. The faster this rate of change is, the stronger is the induced EMF. The change in flux can be made in any number of ways:

- Increasing/decreasing the field strength

- Reversing the direction of the field

- Changing the angle of the field entering the area

- Rotating the area itself

- Moving the area in and out of the field

In short, any mechanism that changes the flux induces an EMF.

## THINGS TO THINK ABOUT

With a modern viewpoint of the importance of symmetry in physics, we can see that this induction of electric field from changing magnetic flux is not so surprising. After all, the magnetic field itself is created only when charges are in motion (and hence their electric fields are changes).

> Changing magnetic fields induce electric fields.
>
> Changing electric fields induce magnetic fields.

A few years later, in 1834, Emil Lenz defined the direction of the induced electric field and, hence, the direction of the induced current if any free charges can move in response to the induced field. His definition of the induced electric field is based on the type of change in flux: the induced EMF opposes the change in flux. For example, if the magnetic flux is increasing, the induced electric fields will be such that they will drive charges to create a new magnetic field in the opposite direction, as shown in Figure 10.2. This is called Lenz's law.

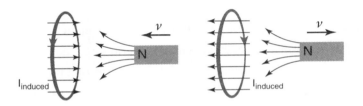

**FIGURE 10.2** Lenz's law.

Traditionally, Lenz's law is indicated by adding a negative sign to Faraday's law:

$$V_{\text{induced}} = \text{EMF} = -\frac{\Delta \Phi_{\text{B}}}{\Delta t}$$

The reason for this opposition is conservation of energy. Recall that all electromagnetic energy is in the fields themselves. If you are moving the magnet yourself in the above cases, you will feel a resistance to your motion. Since you are doing additional positive work to the system in changing the flux, that increase in energy must manifest somehow in the induced electromagnetic fields. Induced fields opposing the change ensure that the external forces are doing positive work to the system.

# 10.2  Common Electromagnetic Devices

Many modern devices use a combination of permanent magnets and electromagnets. The inside of every speaker contains a combination of these two, as shown in Figure 10.3.

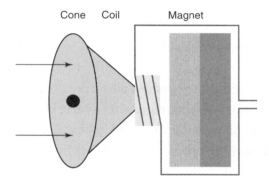

**FIGURE 10.3** A common electromagnetic device
containing a permanent magnet and an electromagnet.

The sound waves from the speaker are generated by the magnetic forces between the electromagnet and the permanent magnet. This magnetic force moves the cone, which compresses or decompresses the air in front of it. By controlling the direction and amount of current going to the electromagnetic coil, the direction and strength of the magnetic force is dictated. A microphone is the same basic device, except the incoming pressure waves move the coil. Since the coil is in a magnetic field, the magnetic flux changes and the induced current produced via induction is recorded!

The electric motor is one of the most useful magnetic devices ever invented. Like the speaker described above, every motor consists of a magnet and a coil of wire. Instead of pushing and pulling, however, the motor rotates.

Figure 10.4 pictures one loop of a coil of wire that is free to rotate inside of a magnetic field. If a current is sent through the wire, it will experience magnetic forces, as shown above. In this way, every motor does its job with a rotating shaft. It is this rotational motion that is at the heart of every electric motor. The rotating coil of wire is called the **armature** of the motor and is shown in Figure 10.5.

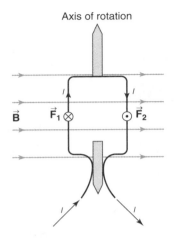

**FIGURE 10.4** An electric motor.

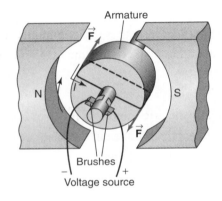

**FIGURE 10.5** The armature of a motor.

The electric grids that deliver energy across the world are all based on alternating current (AC). One of the primary reasons for this is that when the voltage and current alternate, they create changing magnetic fields. Since the changing magnetic field constantly changes the magnetic flux in a nearby coil, the voltage can easily be stepped up (increased) or stepped down (decreased) by changing the relative ratio of the windings on either side of a **transformer** (see Figure 10.6). In the following formula, $N$ stands for the number of windings of wire:

$$\frac{V_{\text{primary}}}{N_{\text{primary}}} = \frac{V_{\text{secondary}}}{N_{\text{secondary}}}$$

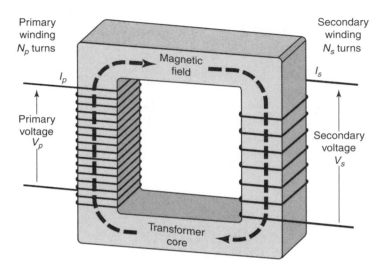

**FIGURE 10.6** A transformer.

Although a magnetic core is not required, they are commonly used to increase the strength of the magnetic field and to help confine the magnetic fields generated to stay within the transformer.

Electric generators are basically electric motors in reverse. Instead of sending electric power in and getting mechanical power out as in the case of an electric motor, an electric generator has its coil of wire turned by external mechanical work and sends out the induced current and voltage as electric power, as shown in Figure 10.7.

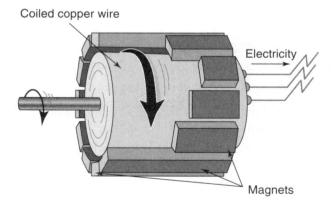

**FIGURE 10.7** An electric generator.

# 10.3 Maxwell and Electromagnetic Radiation

In 1862, James Clerk Maxwell formalized, completed, and summarized the interdependencies of electric and magnetic fields with the charges that produce them into a discrete set of mathematical relationships that became known as Maxwell's equations. At that time, the full symmetry of the fields was finally established. Not only do changing magnetic fields induce electric fields, but changing electric fields induce magnetic fields. In fact, this is the modern understanding of why magnetic fields are sourced by *moving* charges: it is the change in the electric fields around the moving charges that is the actual source of the magnetic fields. As Maxwell was manipulating his equations, he demonstrated that these changing electric and magnetic fields could form a wave traveling through space with a speed related to permittivity and the permeability constant of electric and magnetic fields:

$$\text{Speed of electromagnetic waves} = \frac{1}{\sqrt{\varepsilon_0 \mu_0}}$$

All that is needed to source these electromagnetic waves is an *accelerating* charge. Once created, the changing electric and magnetic fields will self-propagate with no need of a medium. The direction of the wave's motion, the oscillating electric field, and the oscillating magnetic field are all at right angles to each other, as shown in Figure 10.8.

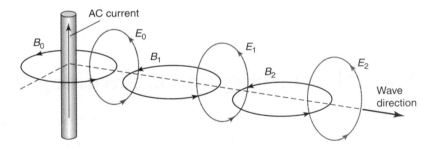

**FIGURE 10.8** An accelerating charge and self-propagating electric and magnetic fields.

Just as an accelerated charge creates the electromagnetic wave, the wave may be absorbed by causing a charge to be accelerated by the electric field of the wave. In this way, energy is transmitted from the source charge to the receiving charge via electromagnetic radiation.

# 10.4 Electromagnetic Spectrum

The frequency of the oscillating electric and magnetic fields of electromagnetic radiation is set by the energy given to the wave by the source charge. There is no restriction on the range of possible frequencies that can be produced. Thus one of the most unifying charts in all of science can now be understood: the electromagnetic spectrum (see Figure 10.9). From the low-energy/low-frequency electromagnetic radiation known as radio waves all the way through to the high-energy/high-frequency electromagnetic radiation known as gamma rays, the entire spectrum of seemingly different phenomena are all just different frequencies of electromagnetic waves traveling through space at the speed of light. The visible light we use to see and its various colors are all just a small sliver of this spectrum. Indeed, the different colors of light are simply different frequencies of electromagnetic radiation.

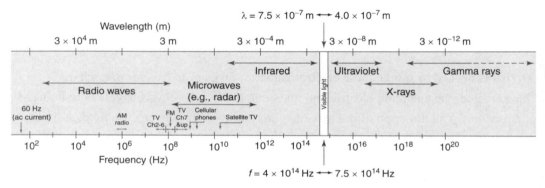

**FIGURE 10.9** The electromagnetic spectrum.

# 10.5 Special Relativity and Electromagnetism

In 1905, Albert Einstein explained the interdependencies of the electric and magnetic fields in his theory of special relativity. Although often presented as a treatise on motion, the speed of light, time dilation, and length contraction, he was actually primarily driven to these conclusions by trying to understand the symmetries of magnetism and electricity. His conclusion is as startling today as it was back then: magnetism and electricity are not just related, they are the same thing! The appearance of the electromagnetic field as either magnetism or electricity or as some combination is due to the relative motion of the source charges and the observer. The electromagnetic force can be categorized and explained consistently as being caused by either a magnetic or an electric field, depending on the reference frame chosen. It is precisely this emphasis on relative motion that explains how and why the change in flux in Faraday's law has the same results whether it is caused by moving the source or by moving the loop.

# Chapter Review Exercises

1. The electric grid routinely steps up and steps down the voltage from hundreds of thousands of volts to 240 volts to send electric power long distances and then into your house. Here's a picture of the transformers routinely placed on residential roads to step the voltage down from the 7,200 V in the main power line.

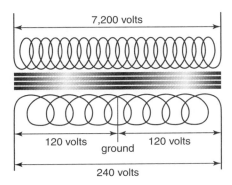

Determine the ratio of windings on the 7,200-volt side of the transformer to the windings on the 240-volt side. Also explain how the typical residence has both a 240- and a 120-voltage line in the house.

2. The conducting rod *ab* shown below makes contact with metal rails *ca* and *db*. The apparatus is in a uniform magnetic field of 0.800 T, perpendicular to the plane of the figure.

   a. Find the magnitude of the EMF induced in the rod when the rod is moved toward the right with speed 7.50 m/s.

   b. In what direction does the current flow in the rod?

   c. If the resistance of circuit *abdc* is 1.50 Ω (assumed to be constant), find the force (magnitude and direction) required to keep the rod moving to the right with a constant speed of 7.50 m/s. You can ignore friction.

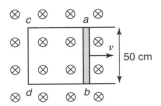

3.   What units do you get when multiplying wavelength and frequency of a wave? What value do you get for any particular type of electromagnetic radiation in a vacuum?

# WAVES

## WHAT YOU WILL LEARN

- Types of waves and their properties.

- Special properties of sound and light waves.

- How the polarized nature of light is used.

- The Doppler shift in sound and light.

- How all types of wave reflect, refract, diffract, scatter, disperse, and interfere.

## 11.1 Simple Harmonic Motion

If the motion of an object is such that it is vibrating or oscillating about an equilibrium point in a sinusoidal fashion (i.e., like a sine or cosine wave), it is said to be undergoing

**simple harmonic motion (SHM)**. What kind of force law causes this type of motion? Any force or sum of forces that can be reduced to look like Hooke's law causes SHM:

$$\vec{F} = -k\vec{x}$$

In this formula, $k$ is a constant determined by the system (often referred to as the **spring constant** since springs commonly obey Hooke's law), $F$ is the net force, and $x$ is the displacement from equilibrium. The negative sign indicates this is a restorative force. That is, the force always acts in the opposite direction of the displacement so as to push the object back toward its equilibrium position. The maximum displacement from equilibrium is called the **amplitude** ($A$) and the time to complete one complete oscillation is the **period** ($T$). Since the unit for period is seconds/cycle, the inverse is cycles/second and is called the frequency ($f$). The unit of cycles/second has its own name: hertz (Hz). Recall that the **angular frequency** ($\omega$) is simply the regular frequency times the number of radians in one complete cycle: $2\pi$. The following formulas relate angular frequency and frequency and relate period and frequency:

$$\omega = 2\pi f$$

and

$$T = \frac{1}{f}$$

Mathematicians prefer to write the force equation this way for reasons that will become apparent shortly:

$$a = -\frac{kx}{m}$$

$$a = -\omega^2 x$$

In this equation, Hooke's law is shown by:

$$\omega = \sqrt{\frac{k}{m}}$$

If we graph $x$, $v$, and $a$ as a function of time, we find that they all are sinusoidal functions—either sine or cosine depending on the initial conditions. The following equations are the standard ways to write the solutions to this general equation:

$$x(t) = A\cos(\omega t + \Phi)$$
$$v(t) = -A\omega\sin(\omega t + \Phi)$$
$$a(t) = -A\omega^2 \cos(\omega t + \Phi)$$

In these standard equations, $A$ is the amplitude of oscillation, $\omega$ is the angular frequency of the motion, and $\Phi$ is the phase shift.

Note that the amplitude $A$ and the phase shift $\Phi$ are determined by the initial conditions. Amplitude is determined by the amount of energy placed into the oscillator initially, either by giving the object a velocity (kinetic energy) or by actively displacing (potential energy) and then releasing it. The phase shift, $\Phi$, is adjusted so that the cosine function has the appropriate value at $t = 0$.

**EXAMPLE**  If a spring is put into motion by stretching the spring a distance $x_0$ and then releasing it, what are the amplitude and phase shift?

**SOLUTION**

$A = x_0$ and $\Phi = 0$ since you need cosine to equal 1 at $t = 0$.

**EXAMPLE**  If a spring is put into motion by giving the spring an initial velocity of $v_0$, what are the amplitude and phase shift?

**SOLUTION**

Set the initial energy equal to the energy at maximum displacement:

$$\frac{1}{2}mv_0^2 = \frac{1}{2}kA^2$$

Conservation of energy gives:

$$A = \sqrt{\frac{m}{k}}v_0$$

You now require cosine $= 0$ when $t = 0$, so $\Phi = \dfrac{\pi}{2}$.

Once you have determined $A$ and $\Phi$ from the initial conditions, you have the complete solutions for displacement, velocity, and acceleration at all times. Note that the angular frequency (and thus the frequency and period as well) are set by the spring constant $k$ and by the mass $m$. They are *not* dependent on the initial conditions.

Although Hooke's law is presented in the context of springs, many objects behave approximately like Hooke's law. Thus the analysis above can be applied. The most common example is that of a simple **pendulum**. It can be shown that for small angles

($\theta < 15°$), the gravitation force that causes the pendulum to move back and forth simplifies to the following:

$$F = -\frac{mg}{L} x$$

In this formula, $L$ is the length of the pendulum and $x$ is its displacement from equilibrium.

By rearranging into the standard form, we find that the angular frequency of any pendulum on Earth is simply $\sqrt{\frac{g}{L}}$, thus making pendulums good timekeepers! No matter what the mass and the initial conditions are (as long as the angle is kept small), the period of oscillation will be the same. All of the above discussions for Hooke's law hold true with $\omega = \sqrt{\frac{k}{L}}$.

At first, it may seem that simple harmonic motion is just an interesting phenomenon in these two isolated cases of springs and small-angle pendulums. However, many, many phenomena can be approximated as simple harmonic motion. To visualize this, consider the difference between a stable equilibrium point and an unstable one. An equilibrium point is a position at which all the forces sum to zero on an object. An **unstable equilibrium point** is one where any slight motion (*perturbation* is the fancy word physicists prefer) will cause the object to move away from the equilibrium point. For example, picture a ball balanced on a hilltop. A **stable equilibrium point** is one where any small perturbation will cause the object to be directed back to the equilibrium point. For example, picture a ball in a narrow valley. Almost any stable equilibrium in any system can be reduced to a Hooke's law situation for tiny displacements. Thus simple harmonic motion turns out to be a very useful approximation under a wide variety of conditions.

Hooke's law represents another conservative force in the context of energy conservation. Since the force is not constant as the spring is stretched or compressed, the work done must be found by calculating the area under the curve of a force vs. displacement graph, where $U_s$ is the potential energy of the spring:

$$U_s = \frac{1}{2} k x^2$$

## 11.2 Wave Basics

A wave is a periodic disturbance of a **medium**, such as air, water, or rope.

If a wave travels through space, it is said to *propagate* and is called a *traveling wave*. Examples include waves in the ocean, sounds, light, and a Slinky in motion.

The points of the maximum positive displacement caused by a wave are called the **peaks**. The opposite points, which are those of the most negative displacement, are called **troughs**. One complete back and forth motion of a wave (including one peak and one trough) is called an **oscillation** or a **cycle**. See Table 11.1 and Figure 11.1 for descriptions of the major attributes of waves.

**TABLE 11.1  SOME MAJOR ATTRIBUTES USED TO DESCRIBE ALL WAVES**

| | |
|---|---|
| **Amplitude ($A$)** | The maximum value of the disturbance, e.g., the height above sea level of an ocean wave |
| **Wavelength ($\lambda$)** | The length of a complete cycle of a wave or the distance between peaks, e.g., the distance from the top of one ocean wave to the top of the next wave |
| **Period ($T$)** | The amount of time for the wave to complete an oscillation, generally expressed in seconds |
| **Frequency ($f$)** | The number of oscillations completed in one second; 1 cycle per second is given the standard unit of frequency: 1 hertz (Hz) |

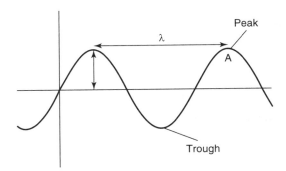

**Figure 11.1** Some major attributes used to describe all waves.

Frequency and period are related as follows:

$$f = \frac{1}{T}$$

If a wave oscillates in a direction at right angles to its propagation, it is classified as a **transverse wave**. A water wave is an example of a transverse wave because it oscillates above and below sea level while it travels along the surface of the water.

If a wave oscillates in the same direction as it propagates, it is called a **longitudinal wave**. Sound is an example of a longitudinal wave because the pressure wave of molecules goes back and forth between the source and the receiver, just like the wave itself.

# 11.3 Traveling Waves

The speed ($v$) at which a disturbance is traveling is called the *wave speed*. Generally, this is pictured as the speed at which the peak of the wave is traveling through space. The distance traveled by one wave in 1 second is the wave speed, which is clearly the wavelength (length per cycle) times its frequency (cycles per second):

$$v = f\lambda$$

A wave's speed depends on its medium, which is the material through which it is traveling. The frequency of a wave depends on the source of the disturbance; the frequency of the oscillating source is the same as the frequency of the wave it generates. When a wave enters a new medium, its frequency does not change. However, the wave may have a new wave speed that would force a corresponding change in its wavelength.

A wave on a string has a speed that depends on the tension in the string ($F_T$ in Newtons) and the string's mass per unit length ($\mu$ in kg/m):

$$v^2 = \frac{F_T}{\mu}$$

If the wave's direction of travel is defined to be the $x$-direction and if the amplitude is in the $y$-direction, we can describe the wave as follows:

$$y(x,t) = A\sin\left(2\pi f t - \frac{2\pi x}{\lambda}\right)$$

Note that in this equation:

- $2\pi f = \omega = $ **angular velocity** in radians/second

- $\dfrac{2\pi}{\lambda} = \kappa = $ **wave number** in radians/meter

By using these identities, the above equation can be simplified to the following:

$$y(x,t) = A \sin(\omega t - \kappa x)$$

# 11.4 Interference

When two waves meet (or overlap), they are said to *interfere* (see Figure 11.2). To determine the resultant of two waves that interfere, simply add their amplitudes, being careful of the positive and negative values. This resultant is called the **superposition** of the original waves.

If, at a certain point, the waves' amplitudes add up to a larger value, the original waves are interfering **constructively**. If, at a given point, the waves' amplitudes add up to a smaller value (one value is negative while the other is positive), the original waves are interfering **destructively**.

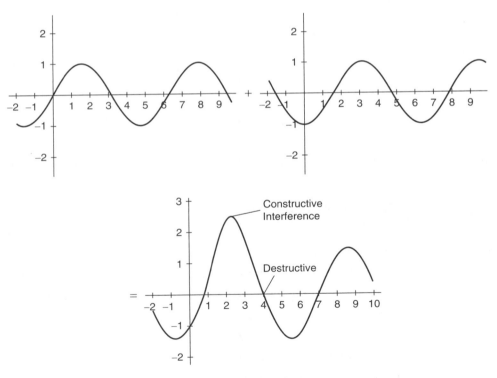

Figure 11.2 Constructive and destructive interference.

If two waves have their peaks aligned, they are said to be "in phase." If one's peak matches up with the other's trough, they are completely "out of phase." (See Figure 11.3.) The phase of a single wave refers to a specific position along its cycle, defined to be zero at its beginning and to be $2\pi$ at its end.

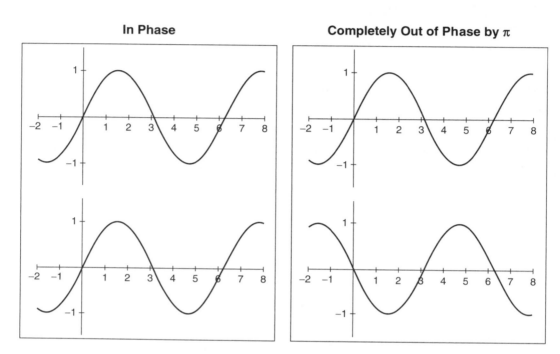

**Figure 11.3** Waves in phase and completely out of phase.

Generally, waves return to their original waveforms after they pass through each other.

# 11.5 Standing Waves

If two repeating waves are overlapping in a region of space, there will be places and times of both destructive interference and constructive interference. If the two waves have the same frequency, the resulting interference pattern could be stationary in space. This is most commonly produced by reflecting one continuous wave back onto itself. The two opposing waves would produce a *standing wave*. The points of complete constructive interference are called **antinodes** (points of widest oscillations), and the completely destructive points are **nodes** (motionless points).

If the ends of the standing wave are forced to be stationary (e.g., the ends of a string are fixed or the ends of a sound tube are closed), the endpoints are nodes. If one end of a string is free to move or one end of a sound tube is open, that point is an antinode.

The distance from one node to the next is half of the wavelength of the standing wave. For fixed endpoints, the shortest possible standing wave contains one antinode surrounded by two nodes at the endpoints. This condition is known as the **first harmonic** or the **fundamental harmonic** ($\lambda = 2L$, where $L$ is the length of standing wave pattern). The number of antinodes in a standing wave is the number of the harmonic that describes the standing wave ($\lambda = \dfrac{2L}{N}$, where $N$ is the number of antinodes).

**Overtones** are the harmonics above the fundamental harmonic. Thus the second harmonic is the first overtone, the fifth harmonic is the fourth overtone, etc. All overtone frequencies are an integral multiple of the fundamental frequency. In general, one standing wave may contain a series of higher harmonics within the principal harmonic you hear or see. This distribution of frequencies is what gives the unique characteristic to two otherwise similar notes (this is known as the *timbre* of a sound). For instance, compare a middle C note on a piano to one on a trumpet. The primary frequency of the note may be the same, but we can still tell the sounds apart because of the higher harmonics.

# 11.6 Sound as a Wave

Sound is a longitudinal pressure wave. Because sound is an especially important wave to humans, we have developed specialized language for describing it. Instead of frequency, we refer to the **pitch** of a sound or note. We refer to sounds of greater amplitude as being louder. Sound amplitude is characterized by its intensity, measured in energy flow per area. We generally use the **decibel** scale to describe the range of intensities of sound that the human ear can detect:

- Threshold of human hearing $= I_0 = 10^{-12}$ watts/m$^2$

- dB (decibels) $= 10 \log\left(\dfrac{I}{I_0}\right)$, where $I =$ intensity of sound

Most humans can hear sounds in the range of 0–120 dB before the loudness becomes painful. We can hear pitches in the range of 20–20,000 Hz. Note that the unit of intensity describes the rate of energy flow per unit area. This means the intensity of the sound wave as it propagates outward from the source will fall as $\dfrac{1}{r^2}$, where $r$ is the distance from the source. Imagine the energy localized on the surface of a sphere going outward and thus the same energy getting spread thinner and thinner over a larger spherical surface.

The ear detects waves of pressure traveling through the air. The source of sound is a vibrating structure (e.g., vocal cords) that perturbs the air next to it at the same frequency of vibration. This causes areas of **compression** (peaks that are more dense) and **rarefaction** (troughs that are less dense) in the medium (such as air, water, or a solid) that travel through the medium. Sound is thus a longitudinal wave that requires a medium of molecules through which to travel. Generally speaking, sounds travel fastest through solids, slower through liquids, slowest through gas (~340 m/s in air), and not at all through a vacuum.

Musical instruments are generally constructed to produce standing waves at certain pitches that correspond to precise frequencies that we call notes. For example, middle A or "la" is 440 Hz, and the next higher note B ("ti") is 495. After the note G ("so"), the next note is the precise double of middle A, 880 Hz. So we cycle through the scale again, starting over with A. This note with doubled frequency that sounds similar to the original note is an "octave" (8 notes) higher in pitch.

If two notes of similar but not identical frequency are played simultaneously, our ears will detect a regular pattern of constructive interference when their peaks overlap in time. This is known as a beat frequency and is equal to the difference between the two frequencies.

If a material settles into a specific frequency of vibration for a wide range of perturbations or disturbances, that frequency is called the material's **natural frequency**. **Resonance** is a general wave phenomenon that occurs when a wave strikes a material at its natural frequency. This causes the material to vibrate at ever-larger amplitudes, and the results can be quite dramatic as the energy of the wave is continually absorbed. Picture an opera singer shattering a crystal glass with her voice.

# 11.7 Light as a Wave

Another wave of critical importance to humans is light. Light is a transverse wave made of two simultaneous waves. One is an oscillating electric field, and the other is an oscillating magnetic field. For this reason, light is more generally known as electromagnetic radiation. They are in phase with each other. Both fields are at right angles to the direction of the propagation and at right angles to each other. For example, if the light beam is going in the $z$-direction, the electric field is going back and forth in the $x$-direction while the magnetic field is oscillating in the $y$-direction.

What humans think of as "light" is more properly termed *visible light,* as we see only a narrow band of frequencies of the infinite possibilities of electromagnetic radiation. Humans can see light between $10^{14}$ and $10^{15}$ Hz with corresponding wavelengths in the hundreds of nm ($10^{-9}$). The frequency in this case refers to the frequency of the oscillating electric and magnetic fields. The entire band of electromagnetic frequencies is known as the **electromagnetic spectrum** and is shown in Figure 11.4.

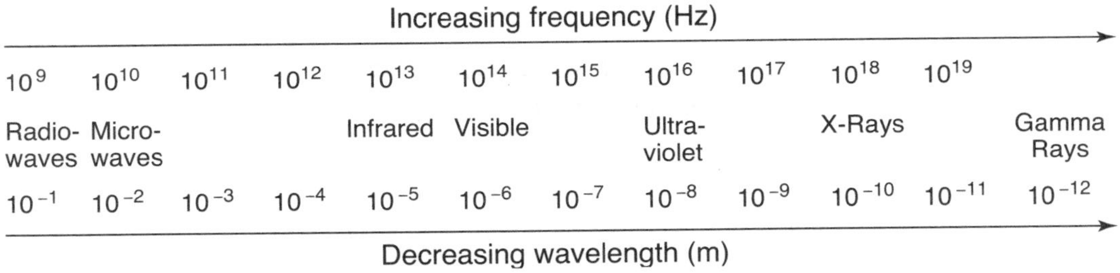

Increasing frequency (Hz)

| $10^9$ | $10^{10}$ | $10^{11}$ | $10^{12}$ | $10^{13}$ | $10^{14}$ | $10^{15}$ | $10^{16}$ | $10^{17}$ | $10^{18}$ | $10^{19}$ |
|---|---|---|---|---|---|---|---|---|---|---|
| Radio- waves | Micro- waves | | | Infrared | Visible | | Ultra- violet | | X-Rays | | Gamma Rays |
| $10^{-1}$ | $10^{-2}$ | $10^{-3}$ | $10^{-4}$ | $10^{-5}$ | $10^{-6}$ | $10^{-7}$ | $10^{-8}$ | $10^{-9}$ | $10^{-10}$ | $10^{-11}$ | $10^{-12}$ |

Decreasing wavelength (m)

Figure 11.4 The electromagnetic spectrum.

All electromagnetic radiation is created by an accelerating charge. The type of charge movement creates light of different frequencies. Although there are many ways of creating electromagnetic radiation, some examples are shown in Table 11.2.

TABLE 11.2 TYPE OF CHARGE AND TYPE OF LIGHT

| Type of Charge Movement | Type of Light |
|---|---|
| Oscillating electrons in large antennae | Radio waves |
| Oscillating charges in molecular (thermal) motion | Infrared |
| Electrons jumping between orbital levels in an atom | Visible |
| Protons shifting in the nucleus of an atom | Gamma rays |

Since light is made of electric and magnetic fields, once it is created by an accelerating charge, it does not need a medium. All electromagnetic radiation travels at the same speed in a vacuum: $3.0 \times 10^8$ m/s. Once light enters a medium, it will slow down by an amount that depends both on the medium and on the light's frequency. In most cases, one can approximate the speed of light to be the same in air as that in a vacuum. To accommodate this change in speed, the wavelength changes in the wave equation. The frequency, however, remains set by the source.

Light, being a transverse wave, has an amplitude at right angles to the direction of motion. For all electromagnetic radiation, this orientation of the electric field oscillation is specifically known as the **polarization** of the wave. (See Chapter 12 for more about polarization.) Often the wave nature of light can be ignored in favor of a raylike model. This field of study is known as optics and is covered in Chapter 12.

 **THINGS TO THINK ABOUT**

Table 11.3 shows how humans perceive the amplitude and frequency of sound and light.

**TABLE 11.3  HUMAN PERCEPTION OF SOUND AND LIGHT**

|  | Sound | Light |
|---|---|---|
| Amplitude | Loudness | Brightness |
| Frequency | Pitch | Color |

# 11.8 Traveling Sources and Receivers: The Doppler Shift

Although waves do not change frequencies as they enter new media, there can be an *apparent* shift in frequencies if either the source of the waves is moving toward/away from the receiver or if the receiver is in motion toward/away from the source. For example, an ambulance siren will dramatically change its pitch as it approaches a stationary listener and as it moves away from the listener. Note, however, that the driver will hear the same constant pitch of the siren. In general, if the source and the receiver are approaching each other, the frequency appears to be higher to the recipient than to the source. If the source and the receiver are getting farther apart, the frequency appears to be lower. This phenomenon is known as the **Doppler shift** and can be described by this equation:

$$f_R = \frac{(v + v_R)f_S}{v + v_S}$$

In the equation above:

- $v_R$ = velocity of receiver

- $v_S$ = velocity of source

- $v$ = velocity of wave

- $f_R$ = frequency of sound heard by receiver

- $f_S$ = frequency of sound heard by source

Note that the sign (+ or −) is chosen such that the receiver's frequency goes up when the velocity is bringing the receiver and source closer, and the receiver's frequency goes down when the velocity is making them farther apart.

Visible light undergoes Doppler shifts as well.

If an object emitting electromagnetic radiation is approaching us, we say its light is *blue-shifted*. If the object is moving away from us, we refer to its light as being *red-shifted*.

Shock waves can be produced by waves when the source is traveling faster than the speed of the wave in that particular medium. In the case of sound waves, this is the origin of the sonic boom (see below for other examples).

# 11.9 Wave Behavior

As waves travel and encounter obstacles and changes in mediums, they will change their usual patterns by reflecting, refracting, diffracting, scattering, and dispersing. Although some specifics regarding the reflection, refraction, and diffraction of light waves are covered in a subsequent chapter (Chapter 12), all waves experience these phenomena. Because of the vastly different wavelengths of sound compared with those of light, it is easier to observe some of these phenomena in everyday life with sound waves, light waves, or mechanical waves. However, all waves exhibit all of these behaviors under the appropriate conditions! An important principle in understanding some of the behavior of traveling waves is known as Huygens principle. In the model, each new wave front of the traveling wave is sourced by the previous wave front of the same wave. Often, whether or not the phenomenon is commonly experienced depends simply on the wavelength involved in our daily interactions with sound, light, and mechanical waves. See Table 11.4.

### TABLE 11.4  EXAMPLES OF WAVE BEHAVIOR

| Wave Behavior | Sound Example | Light Example | Mechanical Example |
|---|---|---|---|
| **Reflection:** As the energy of the wave enters the new medium with a different wave speed, the wave will partly be sent back into the original medium. | **Sonar/echolocation:** Bounce a sound wave off of a surface and measure the time until you detect the echo. By knowing the speed of sound and time, you calculate distance. | **Mirrors:** You see the light originally from your face going to the mirror, bouncing off, and then going into your eyes. | **Water waves:** The wave from a diver entering a pool bounces off of the sides of the pool. |
| **Refraction:** In order to satisfy continuity conditions, as the wave continues into a new medium, it may propagate in a different direction. | **Sounds over a cool pond:** Sound travels slower in the cool air just above the water, so the sound waves that would normally be "lost" to the air even higher up refract back into the cooler air, increasing the intensity of sound carried across the pond. | **Glasses:** Lenses refract the light from air to glass such that they are redirected in a way that your eye can now bring them into focus (by refracting with your natural lenses). | **Shallow and deep water:** Water waves travel at different speeds depending on the depth, so they appear to bend and warp as they travel over shallows. |
| **Scattering:** If the obstacles encountered by the waves are comparable in size to the wavelength of the wave itself, the wave may be reflected in all directions. | **Sound tiles behind stages:** In order to preserve sound energy into the audience but not to cause a distinct echo, multiple curved surfaces are employed. | **Blue skies:** Nitrogen molecules in the atmosphere scatter the shorter blue wavelengths within the white light from the sun more than wavelengths of other colors. | **Earthquakes:** As the wave travels through the ground, it scatters in all directions when encountering substantial mineral deposits. |

| Wave Behavior | Sound Example | Light Example | Mechanical Example |
|---|---|---|---|
| **Diffraction:** The edges of obstacles or openings act as sources for the wave on the other side. The relative size of the wavelength and the objects/openings must be comparable. | **Open doors:** You can hear people from around corners and through open doorways even when you cannot see them since sound waves detected by your ears are much longer than light waves detected by your eyes. | **Ray of light:** A tiny hole in a curtain lets in a ray of light that grows in size until it strikes the far wall as a large disk. | **Ocean waves:** When ocean waves enter a harbor through its narrow opening, the wave spreads to all corners of the harbor. |
| **Interference:** Two waves superimposed can make a bigger wave (constructive interference) or a smaller one (destructive interference). | **Noise-canceling headphones:** An embedded microphone detects incoming sounds and then an oppositely phased sound wave is produced by the embedded speakers such that it destructively interferes with the original sound. | **Thin-film interference:** Visible light reflects off of the top surface of an oil layer on top of water, which interferes with the same light reflecting off of the bottom of the oil layer. Depending on the thickness of the layer, these two rays may experience interference. | **Water waves:** As the wake from a boat passes through the ocean waves, an extralarge wave (constructive) or a temporary calm wave (destructive) will result. |
| **Shock wave:** If the source of the wave is moving faster than the speed of the wave itself, then a trailing, conical shock wave will be generated. | **Sonic boom:** An airplane exceeding the speed of sound in air produces a large pressure wave in the air. | **Cerenkov radiation:** Charged particles can go faster than the speed of light in a nonvacuum medium (like the air); when they do, blue light is emitted. | **Water wake:** Boats or animals traveling faster than the speed of the water waves create a large-amplitude wave. |
| **Standing waves:** A regular pattern in space of constructive and destructive interference between two waves of the same frequency travelling in opposite directions. | **Blowing over a bottle top opening:** The frequency of sound created corresponds to the wavelength of the standing wave formed inside the bottle. | **Michelson interferometer:** By splitting a single beam of light into two paths and then overlapping them with mirrors, a standing light wave is created. | **Guitar strings:** All musical instruments create a certain pitch by the creation of standing waves within. By changing the length of the string, the wavelength of the standing wave is adjusted. |

| Wave Behavior | Sound Example | Light Example | Mechanical Example |
|---|---|---|---|
| **Resonance:** Large amplitudes can be built up by absorbing energy at the natural frequency of the object. | **Shattering glass with your voice:** If the glass has a natural frequency of vibration, it can be shattered by matching its pitch with a large-amplitude (loud) sound wave. | **Microwave oven:** By reflecting the single wavelength of electromagnetic radiation within the microwave, cold (destructive) and hot (constructive) spots are created. | **Jump rope:** One end is held still, and the other end is oscillated at the frequency that corresponds to a wavelength of the separation. |
| **Doppler effect:** The wavelength and frequency of a wave apparently change due to the relative motion of the source. | **Car sounds:** As a car passes you, the pitch of its engine goes from high (approaching) to low (receding). | **Radar gun:** The police officer bounces a radio signal off of your car. The reflected, Doppler-shifted wave is compared to the original wave to determine the speed of your car. | **Water waves:** If an object is slowly moving through the water, its ripples are spaced more closely in front and farther apart behind. |
| **Dispersion:** A form of refraction in which the speed of propagation is noticeably different for different frequencies of the wave. In this way, the various frequencies in the original wave are separated. | **The cochlea of the ear:** The inner ear separates the sound wave produced by the eardrum by frequency because the speed of the sound waves varies as it goes through the tube (cochlea), which has a changing diameter. | **Rainbows:** As light enters a suspended drop of water (or prism), it refracts at different angles depending on the frequency, thus separating the colors. | **Pulses on a wire:** One pulse is separated into its various frequencies along a wire of changing thickness as the speed of waves along the wire changes. |

# 11.10  Encoding Information in Electromagnetic Waves

The use of electromagnetic waves has proliferated to the point where modern life would be unimaginable without this invisible network transferring information around us all the time at the speed of light. In addition to the waves generated by technology, the scientific community has long used the information encoded in naturally produced electromagnetic waves. All wireless communication is done via the transmission of electromagnetic radiation. The information itself rides along the wave in one of the four attributes of the waves listed below. The transmitter and receiver have an agreed upon "key," or method, of encoding or decoding the information:

- Amplitude—strength of signal

- Frequency—energy of the moving charges in the source

- Phase—timing of the signal emission and detector absorption

- Polarization—relative orientation

There are many, many examples in both technology and the scientific communities of how these four attributes of waves are used. Table 11.5 lists just a few examples.

### TABLE 11.5  TECHNOLOGICAL AND SCIENTIFIC USAGES OF WAVE ATTRIBUTES

|  | Technological Uses | Scientific Uses |
|---|---|---|
| Amplitude variations | AM radio, digital radio | Brightness of source |
| Frequency variations | FM radio | Doppler shifts |
| Phase variations | RFID (radio frequency identification) | Light traveling through plasma or gas becoming phase shifted |
| Polarization variations | Satellite communication: more signals at the same frequency can be sent at once if they have different polarizations | Polarization angle rotating when passing through plasma |

# Chapter Review Exercises

1. a. Draw a picture of the following standing waves in the chart below. Note that the string is 1.35 meters long and the velocity of the wave on the string is 180 m/s.

| | String Held Fixed at Both Ends | Frequency | Wavelength |
|---|---|---|---|
| $N = 1$; 1st harmonic; fundamental frequency | | | |
| $N = 4$; 4th harmonic; 3rd overtone | | | |

   b. If the fundamental frequency of the waves described above produce sound in air, what is the pitch and wavelength of each sound wave generated? Write your answer in the chart above.

   c. One end of the string in the wave described above was allowed to move up and down freely while the other end remained fixed. Redo the chart for the first harmonic.

| | String Held Fixed at One End While Loose at the Other | Frequency | Wavelength |
|---|---|---|---|
| $N = 1$; 1st harmonic; fundamental frequency | | | |

2.  A 780 Hz siren is on an ambulance approaching you at 27.5 m/s while you are standing motionless on the side of the road.

    a.  What pitch will you hear as the ambulance approaches you?

    b.  What pitch will you hear after the ambulance passes you and is moving away?

3.  a. What is the decibel (dB) rating of a sound wave with an intensity of $10^{-7}$ W/m$^2$?

    b. What is the intensity of a 90 dB noise?

4.  A spring of mass 7.5 kg moves as shown in the graph. The $y$-axis has units of centimeters. The $x$-axis has units of seconds. (Ignore the negative time.)

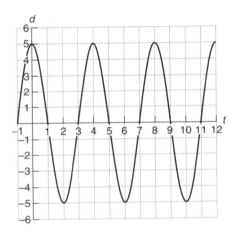

    a.  What is the amplitude?

    b.  What is the period?

    c.  What is the frequency?

    d.  What is the angular frequency?

    e.  What is the spring constant?

    f.  What is the total energy of this oscillator?

    g.  Mark with a circle the first two points of maximum kinetic energy.

    h.  Mark with an "X" the first two points of maximum elastic potential energy.

i.  Determine the maximum force experienced by this mass.

j.  Determine the maximum velocity experienced by this mass.

k.  Write down a complete equation for this motion as a function of time:

$$y(t) =$$

5.  Into which band of the electromagnetic spectrum would a 1.5-picometer wave be placed? What would its wavelength be in a vacuum?

# OPTICS

## WHAT YOU WILL LEARN

- Why mirrors almost always have metal backing.

- How to use Snell's law in refraction problems.

- How fiber-optic cables work.

- How to ray trace light for simple optical systems.

- What spectroscopy is and how it is used.

- Why we sometimes see a rainbow of color in an oily puddle.

- Why the sky is blue.

### LESSONS IN CHAPTER 12

- What Is Optics?
- Law of Reflection
- Law of Refraction
- Ray Tracing and Image Formation

- Two-Slit Interference and Diffraction Gratings
- Single-Slit Interference
- Thin-Film Interference
- Scattering

## 12.1 What Is Optics?

Light is so fundamental to our understanding of the universe that we have devoted an entire field of study to it: optics. For the most part, optics deals with the macroscopic, sometimes referred to as classic, properties of light in situations where light can be modeled as rays. In this ray model of light, the propagation of light is represented by a series of arrows representing the motion of the wave fronts. Optics is mostly concerned with visible light, which has very small wavelengths when compared with most objects in our daily lives. However, the rules of optics do apply to any electromagnetic radiation as long as the wavelength is small enough in the context of the experiment to be modeled as consisting of rays.

## 12.2  Law of Reflection

Reflection is the phenomenon that occurs when light strikes a surface (the **incident** ray) and bounces back (the **reflected** ray). Recall from conservation of momentum that sideways momentum must be preserved during a collision with a surface. If you apply the same thinking to light, you have the **law of reflection** (see Figure 12.1): the **angle of incidence** ($\theta_i$) and **angle of reflection** ($\theta_r$) are equal.

$$\theta_i = \theta_r$$

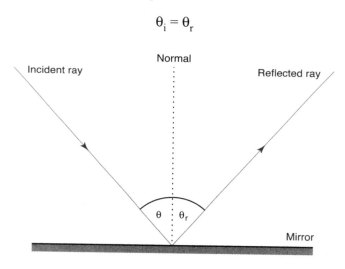

**FIGURE 12.1** The law of reflection.

Note that in optics, angles are measured from the perpendicular (normal) of surfaces.

A smooth surface reflects images, preserving the orientation of the ray. This behavior, which is seen in mirrors, is known as **specular** reflection. A rough surface obeys the law of reflection locally (using the tangent as the line of surface for a curved surface) but does not reflect the image intact. This behavior is known as **diffuse** reflection.

Microscopically, a reflection is the absorption of a photon by a charge and the subsequent reemission of the energy as another photon. If the absorbing charge cannot then vibrate or change energy levels, the photon may not be reemitted and the energy will be absorbed (most likely as thermal energy). Metals tend to be highly reflective because they have the conduction layer electrons that are free to move and because they can easily be manufactured to have flat surfaces. Many materials both reflect some light and allow some to pass through (e.g., glass, water), which brings us to our next topic—refraction.

# 12.3 Law of Refraction

When light enters a new medium, its speed changes. Recall that in a vacuum, all electromagnetic radiation travels at the same speed: $c = 3.00 \times 10^8$ m/s. Any medium besides a vacuum slows down light because that light has to pass through additional electric and magnetic fields due to the existing charges in the medium. These additional fields slow down the progress of the light's fields. In general, the amount by which light is slowed depends on the light's frequency as well as the medium. Frequently, however, it is good enough to associate a single average speed for visible light traveling through the medium. Rather than list the speed of light in the medium, we refer to the medium's **index of refraction** ($n$):

$$n = \frac{c}{v}$$

In the equation above, $v$ is the speed of light in the material.

The minor variation in $n$ for different frequencies of light leads to the dispersion of colors in a prism. The visible spectrum of colors in order of their frequencies is what we call a rainbow.

A ray of light that is directly incident ($\theta_i = 0$) upon a material will slow down at once by shortening its wavelength but not its frequency. That ray of light will continue going forward in a straight line. If the ray is incident at any nonzero angle, however, the ray will "bend" into the slower material, toward the perpendicular on the inside. The ray will then continue in a straight line along its new heading in the medium. This bend caused by the change of indexes of refraction is known as **refraction**. To find the angle of refraction rigorously, we must apply conditions of continuity to the electric and magnetic fields at the surface. However, by simply using the ray model (see Figure 12.2), the law of refraction as stated by Snell determines the relative incident and refracted angles in **Snell's law of refraction**:

$$n_1 \sin\theta_1 = n_2 \sin\theta_2$$

The subscripts 1 and 2 denote the two different media.

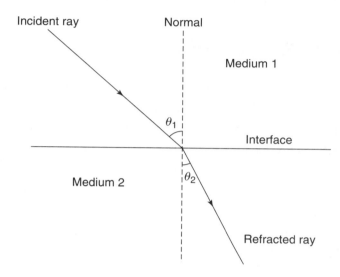

**FIGURE 12.2** Snell's law of refraction.

The argument presented above can be reversed for a ray emerging from a slower medium and entering a faster one. In that case, the ray is bent *away* from the perpendicular in the faster medium. Thus, Snell's law works for any two indexes of refraction.

Or does it? Imagine the case where a ray is going from a slower medium (with a higher value for $n$) to a faster one (with a lower value for $n$) and the angle of incidence is close to parallel to the surface (large $\theta_i$). When solving for the refracted angle in the faster medium, we could get $\sin\theta_r > 1$. This is not possible! There will not be a refracted ray at all, and the incident ray must be entirely reflected to conserve energy. This is known as **total internal reflection**, meaning that no light is refracted out of the slower medium. The limiting case is when the angle is such that $\sin\theta_r = 1$. The incoming ray that meets this condition is called the **critical angle**. In this case, the "refracted" light is forced to stay along the surface of the interface ($\theta_r = 90°$) and does not really emerge into the faster medium. This phenomenon is used every day in fiber-optic cables where light pulses are kept within a single thin fiber without the use of any reflective coating. If $n_1 > n_2$ then the following is true:

$$n_1 \sin\theta_c = n_2 \sin 90° = n_2$$

In this equation, $\theta_c$ gives the largest incident angle from the slower medium that is not 100% reflected.

# 12.4 Ray Tracing and Image Formation

In the cases of thin lenses and mirrors, we can incorporate the details and effects of refraction and reflection into something called the **focal length** ($f$). The focal length of a mirror or lens is the distance at which the lens or mirror causes parallel rays of light to intersect at one point, $f$. **Converging** lenses focus incident light onto a focal point on the opposite side of the lens. Such lenses are sometimes called positive lenses because their focal length is positive. **Diverging** lenses refract the light such that the focal point appears on the incident side of the lens. These lenses are sometimes called negative lenses because their focal length is negative. A **concave** mirror is a diverging lens because it reflects light back to a point at its focal distance on the incident side. In contrast, a **convex** mirror is a converging lens because it reflects the light outward such that the light appears to have come from a focal distance behind the mirror. See Figure 12.3.

By using these simple but powerful concepts, one can simply trace rays (a process known as **ray tracing**) from an object through an optical device to determine where the image will be located. To determine an image's location, draw two out of the following three rays from the top of the object.

- A ray parallel to the central axis that then follows the known reflected or refracted path as shown in Figure 12.3

- A ray that passes through a focal point such that we know its emerging path by reversing one of the diagrams in Figure 12.3

- A ray that passes straight through the center of a lens emerging unrefracted

- A ray that passes straight through the center of a mirror being reflected as if from a plane mirror

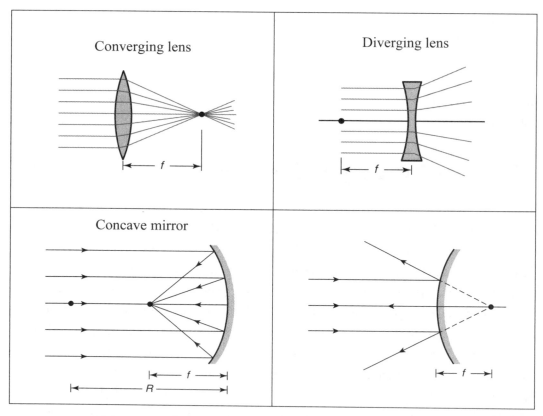

**FIGURE 12.3** The focal point in various lenses and mirrors.

If the emerging rays actually cross, the image is said to be **real**. For example, a camera placed at the point where rays cross record the image on film. If the rays do not intersect but must instead be traced backward to find a point of intersection, the image is said to be **virtual**. A camera placed at a point where rays do cross would not record the image on film since the light rays are not really intersecting there!

If the image is upright (the image's top is on the same side of the axis as the object's top), the image is said to be **upright**. A flipped image is said to be **inverted**. Real images are always inverted, whereas virtual images are always upright in these simple systems. The image can be bigger (**enlarged**) or smaller (**diminished**). Magnification is the ratio of the height of the image to that of the object.

**EXAMPLE** What type of image is created when an object is placed inside the focal length of a converging mirror?

**SOLUTION**

Ray tracing for an object inside the focal length of a converging mirror yields an upright, enlarged, virtual image.

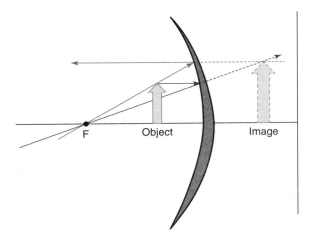

**EXAMPLE** What type of image is created when an object is placed outside the focal length of a converging lens?

**SOLUTION**

Ray tracing for an object outside the focal length of a converging lens yielding a real, inverted, enlarged image.

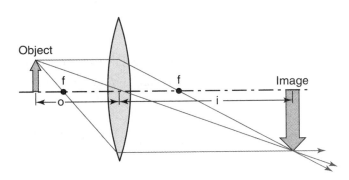

What type of image is created when an object is placed outside the focal length of a diverging lens?

**SOLUTION**

Ray tracing for an object outside the focal length of a diverging lens yielding a virtual, upright, diminished image.

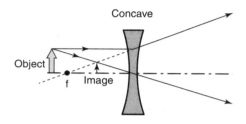

What type of image is created when an object is placed near a diverging mirror?

**SOLUTION**

Ray tracing for any object near a diverging mirror yielding an upright, diminished virtual image.

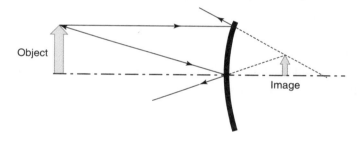

If the distance from the object to the lens or mirror is $d_o$ and if the distance from the lens or mirror to the image is $d_i$, we can use the **lens** or **mirror equation**:

$$\text{Lateral magnification} = \frac{\text{Image height}}{\text{Object height}} = -\frac{d_i}{d_o}$$

Here is some jargon specifically associated with lenses:

$$\text{Power} = \frac{1}{f}$$

Optical power has units of diopters $\left(\dfrac{1}{\text{meter}}\right)$.

Mirrors and lenses are carefully manufactured to get their focal length using these formulas:

$$\text{Radius of curvature} = 2f\,(\text{mirrors})$$

$$\text{The lensmaker's equation:}\ \frac{1}{f} = (N-1)\left(\frac{1}{r_1} + \frac{1}{r_2}\right)$$

In the lensmaker's equation, $n$ is the index of refraction of the material the lens is made from, and $r_1$ and $r_2$ are the two radii of the curves of each side of the lens. See Table 12.1.

**TABLE 12.1  IMAGES FORMED BY VARIOUS LENSES AND MIRRORS**

| | Sign of Focal Length | Object Location | Image Type/Orientation | Image Size |
|---|---|---|---|---|
| Converging systems (convex lenses and concave mirrors) | + | $> 2f$ | Real and inverted | Diminished |
| | | Between $f$ and $2f$ | Real and inverted | Enlarged |
| | | $< f$ | Virtual and upright | Enlarged |
| Diverging systems (concave lenses and convex mirrors) | – | Any | Virtual and upright | Diminished |

# 12.5 Two-Slit Interference and Diffraction Gratings

As we know from studying waves, constructive and destructive interference are unique properties of waves. For example, if the two waves have amplitudes in opposite directions when they meet, their temporary superposition as the waves pass through each other will have a smaller amplitude, which is destructive interference. Although there are many reasons to support the wave model for light, to demonstrate wavelike behavior on the part of light convincingly, we need to show both constructive and destructive interference. **Young's two-slit experiment** is the classic example of this, as shown in Figure 12.4.

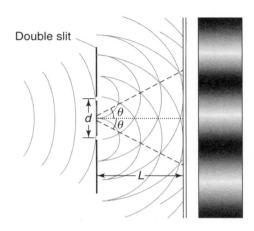

Double slit

**FIGURE 12.4** Young's two-slit interference experiment.

Two narrow slits a distance of *d* apart have monochromatic, in-phase light shining through them onto a screen a distance *L* ($\gg d$) away. If light was made of particles, we would simply expect to see two bright strips (or **fringes**) directly in front of each slit. However, this is not what we see! Instead, a diffraction (or interference) pattern is observed. Each slit diffracts its light in all directions, and every point on the screen receives light from both slits. If the two waves arrive at that point on the screen in phase, a bright fringe is observed. If the two waves arrive at a certain point out of phase, a dark fringe (no light) is observed.

To calculate the relative phase of the two beams of light, we must look at each ray's path from its slit to a specific point on the screen. If the difference in path lengths ($\Delta l$) is an integral multiple of the light's wavelength, there is constructive interference. If the difference is half of the light's wavelength, the beams of light are completely out of phase and there is destructive interference. For bright fringe (constructive interference):

$$\Delta l = m\lambda$$

For dark fringe (destructive interference):

$$\Delta l = \left( m + \frac{1}{2} \right)\lambda$$

In the equations above, $m = \ldots, -3, -2, -1, 0, 1, 2, 3, \ldots$ (any integer).

Using the fact that $d \ll L$, the diffraction pattern for the two-slit experiment has bright fringes at the following angles. Note that angles are measured from the central horizontal axis of the experiment:

bright fringes (*m*th order where *m* is any integer as above)

If we define the *x*-axis to be along the screen with $x = 0$ at the central maximum ($m = 0$) opposite the center of the two slits, this equation can be rewritten as the following:

$$x = \frac{m\lambda L}{d}$$

Note that these formulas give the location of the bright fringes but not their intensity. The intensity is strongest at the central maximum ($m = 0$) and fades quickly and symmetrically to either side. See Figure 12.5.

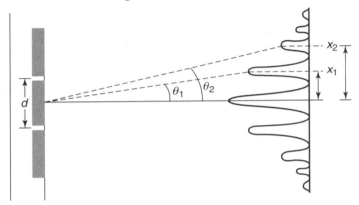

**FIGURE 12.5** Fringe pattern for two-slit interference.

If we add more evenly spaced slits equal distances $d$ above and below the first two, we will not shift the positions of the bright fringes. However, we will tend to even out and maintain the intensities of the bright spots. A large number of evenly spaced slits (or lines) is called a **diffraction grating** and, as might be expected, has the same formula as above. A diffraction grating has the advantage of producing a larger, sharper, and brighter interference pattern. In fact, it is a commonly used tool in science to determine the wavelengths of unknown sources.

A light source containing several different frequencies of light can be resolved into its component frequencies with a diffraction grating. The distinct pattern of intensities as a function of frequency is called a **line spectrum**. Correspondingly, this field of analysis is known as **spectroscopy** and the device used (diffraction grating plus analysis of the interference pattern by frequency) is called a **spectroscope**. These devices play a key role in our understanding of both the very small (the quantum mechanics of atoms) and the very large (the composition of stars and other astronomical objects). Note that the line spectra may be useful both in the light being transmitted (in the case of a few frequencies) or the frequencies that are no longer present (the missing frequencies) in the case of white light passing through an absorbing material. The material absorbs only those frequencies that correspond to its own resonance frequency, and thus scientists can use spectroscopy to tell what something is by what is missing!

## 12.6 Single-Slit Interference

The light from a single slit can also produce an interference pattern, as shown in Figure 12.6. In this case, the many rays of light that pass through a slit of width $D$ form the diffraction pattern. The central position is a bright spot. However, as we move away from the central bright fringe, there is an angle at which the rays all destructively interfere. This can be found by finding the positions of constructive interference from the two-slit experiment since each adjacent ray coming from inside the slit has a partner ray that is out of phase. The following formula can be used to identify the dark fringes for integer $m$ where $m \neq 0$:

$$D\sin\theta = m\lambda$$

The width of the central maximum can be determined by going from $m = -1$ to $m = +1$.

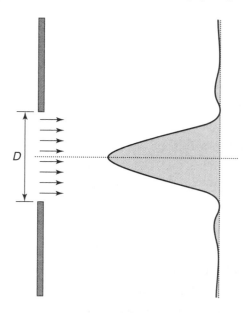

**FIGURE 12.6** Single-slit interference.

# 12.7 Thin-Film Interference

Interference happens whenever two waves overlap. If a wave encounters a thin, transparent film, it will reflect upon entering the film. The transmitted part will reflect again on the opposite side. These two reflected waves will overlap as they exit the thin film back into the side from which the light came. Since one wave traversed the thin film twice, the two waves have different path lengths. Whether constructive or destructive interference occurs depend both on the difference in path length and on the wavelength of the incoming light as explained in the two-slit interference pattern above.

Two things should be mentioned when applying the difference in path length to thin films.

1. The new wavelength of the light $\left(\dfrac{\lambda}{n}\right)$ in the thin film (of refraction index $n$) must be used, as this is where the paths differ.

2. Reflection sometimes causes a phase shift in the wave. When reflecting from a higher to a lower index of refraction, wave phase is preserved during the reflection. When reflecting from a lower to a higher index of refraction, the reflected wave is **phase shifted** by half a wavelength (180° or $\pi$). This can be pictured in terms of a wave on a string encountering a fixed endpoint and bouncing back (phase shifted, like light encountering a slower medium) or a wave encountering an endpoint that is free to move up and down (no phase shift, like light entering a faster medium).

 **THINGS TO THINK ABOUT**

A handy way of remembering when a reflected wave has a phase shift:

"Low to High,
Phase shift pi.
High to Low,
Phase shift no."

**Newton's rings** are a special case of thin-film interference where rings of interference are seen when two plates of glass are placed one upon the other. If the top glass is slightly curved, the varying path lengths in the air gap cause a regular pattern of interference leading to the pattern of rings.

Recall that white is what we call light containing all frequencies. White light incident on a thin film has different constructive interference happening at slightly different angles of viewing due to the component lights' differing wavelengths. (The differing thickness of the thin film also plays a role.) Thus, interference explains the distorted rainbows sometimes seen in a puddle of water when there is a thin film of oil on top.

# 12.8 Scattering

Small particles tend to disrupt the path of small-wavelength light more so than long-wavelength light. Thus, tiny air molecules scatter different frequencies of light differently. The dependence is strongly correlated to wavelength such that the scattering intensity of light is proportional to the fourth power of frequency ($\sim f^4$ or $\sim \frac{1}{\lambda^4}$). Thus the blue light in white light is more strongly scattered. When we "see" the sky not directly in front of the sun, what gets to our eyes is the scattered light. Since blue is more strongly scattered, the sky appears blue! Sunsets appear red for the same reason. Although the visible light from the sun consists of all frequencies (and thus should appear white), by the time the light from the setting sun has gotten to our eyes, so much blue has been scattered by the atmosphere that the sky appears red!

## POLARIZATION

Light, being a transverse wave, has its amplitude at right angles to the direction of motion. For all electromagnetic radiation, this orientation of the electric field oscillation is specifically known as the **polarization** of the wave. Although each individual photon of light has its own polarization, a beam of light or a specific light source may contain or be emitting a random distribution of orientations. This light is called *unpolarized*. A source (such as an LED screen) that emits light of a single type of orientation is called *polarized*. Special filters that allow only certain orientations of light through (by means of embedded, closely spaced conducting material) are called *polarizers*. See Figure 12.7.

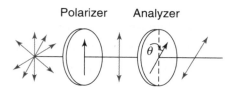

**FIGURE 12.7** Polarizing filters.

Since the intensity of light is proportional to the amplitude squared, the intensity of light that makes it through the second polarizer is reduced for the polarized light incident upon it by cosine squared. This is known as the law of Malus:

$$I_{final} = I_{polarized} \cos^2 \theta$$

Polarization can happen through reflection as well. Imagine unpolarized light incident upon a surface. As the angle of incidence gets larger (the ray is lower to the ground), those rays with electric fields oscillating in the place of the surface are more likely to be reflected. Eventually, the light becomes completely polarized at Brewster's angle ($\theta_p$), as shown in Figure 12.8.

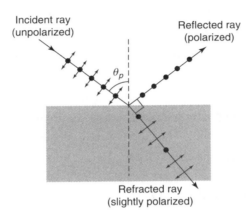

**FIGURE 12.8** Polarization through reflection.

$$\tan \theta_p = \frac{n_2}{n_1}$$

In the equation above, known as **Brewster's law**, $n_1$ is the incident light's medium.

Although our eyes are not sensitive to the polarization of light, certain sunglasses make use of the fact that the process of scattering tends to polarize light. On a bright day, polarized sunglasses preferentially eliminate the light reflected from the ground and scattered from the air, thus making it easier for our eyes to focus on the other objects in our field of view. The phenomenon of polarized light is the best physical example of the transverse nature of light waves.

Another common use of polarized light is at the movie theater. The glasses worn when viewing 3-D movies are made from two different polarizers. One eye sees clockwise-polarized light, while the other eye sees counterclockwise-polarized light. In this way, the movie can direct two different images on the screen to your two different eyes, giving you the illusion of depth that your eyes naturally produce when looking at real, 3-D objects around you from your eyes' two slightly different locations.

# COLOR THEORY

Why have our eyes evolved to respond differently to the frequencies of light that we now call "visible"? Put another way, why do we see colors? These visible frequencies of light are heavily influenced by the energy levels between electron orbitals in molecules and atoms. When we see an object has a certain color, all the other frequencies of the white light were **absorbed** and not reemitted by that object. The emitted light corresponds to a certain frequency given off by an electron that absorbed that frequency of light specifically and hopped to a higher energy level. The **reemission** of that photon of light happens when that stimulated electron jumps back down to its original orbital level. The other frequencies are absorbed as thermal energy that heats up the object. So, in essence, the color tells us something about the chemistry of an object. The ability to sort items by sight based on chemical composition has some clear biological benefits for animals when eating. Survival depends on eating. Thus the ability to differentiate colors has a survival value and is a trait likely to be passed to future generations.

 **THINGS TO THINK ABOUT**

Why do we see only certain colors if the frequency band is continuous? Our eyes have evolved specialized receptors called *cones* that can differentiate among different frequencies of visible light. Most people have three different types of cones, each of which has its own specialized band of frequencies. These three frequency bins are what we mentally sort into the colors we "see" in the continuous spectrum. This is why we see only six or seven bands of color in the continuous spectrum of a rainbow and why the entire color spectrum, as seen by humans, can be re-created from just three primary colors.

**Phosphorescence** is explained by this understanding of electrons. When an object is subjected to light, electrons absorb the energy and jump to a higher orbital. After the stimulation stops (the object is in darkness), the electrons then begin to fall back down one by one, each emitting a photon. Thus the object glows in the dark for a while as the excited electrons fall back down to their lower-energy orbitals.

# Chapter Review Exercises

1. Trace the following 3 objects with at least two rays. Draw and label your image as *i*. Then circle the correct attributes (one from each pair) of the image.

### Converging Lens

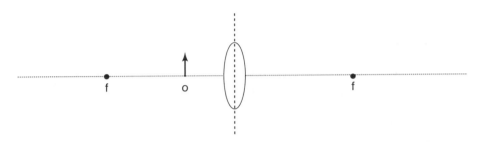

Virtual or Real      Upright or Inverted      Magnified or Diminished

**Convex Mirror** (center of curvature is indicated)

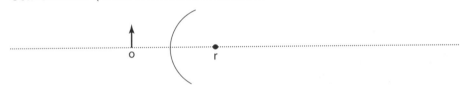

Virtual or Real      Upright or Inverted      Magnified or Diminished

### Diverging Lens

Diverging Lens

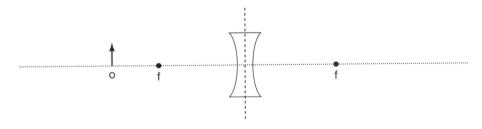

Virtual or Real      Upright or Inverted      Magnified or Diminished

2. A ray of light is incident on an unknown transparent substance from the air. The angle of incidence is observed to be equal to 40°, and the angle of refraction is observed to be equal to 22°.

   a. Calculate the index of refraction for this substance.

   b. Calculate the velocity of light in this substance.

   c. The substance is now submerged in glycerol ($n = 1.47$). Calculate the critical angle of incidence for light going from this substance into glycerol.

   d. What happens in situation (c) if the light coming from the substance into glycerol is at an angle larger than the critical angle?

3. A very thin layer of oil ($n = 1.6$) is on top of a puddle of water ($n = 1.3$). Red laser light (650 nm) is observed to undergo destructive interference (cannot be seen in reflection).

   a. What is the wavelength of light in the oil?

   b. Is there a phase shift due to the reflection between the air to oil surface? If so, what is that phase shift?

   c. Is there a phase shift due to the reflection between the oil to water surface? If so, what is that phase shift?

   d. Why is the path length difference twice the thickness of the oil layer?

   e. What is the minimum nonzero thickness of the layer of oil to produce this destructive interference?

   f. Will the very edges of the oil layer (where the thickness is going to zero) be bright or dark under reflection? Why?

   g. Does your answer to part (f) depend on the wavelength of light used?

4.  Two closely spaced slits produce an interference pattern using a red laser light of 650 nm ($n = 10^{-9}$), on a wall 5 meters away in which the bright spots are 1.0 cm apart.

    a.  What is the distance between the slits?

    b.  What would happen to the spacing between the bright spots if only a green laser light was used?

    c.  What would happen to the spacing between the bright spots if only the spacing between the slits was increased?

    d.  What would happen to the spacing between the bright spots if only the wall was moved back to be farther from the laser and slits?

5.  Why are there no filters for polarized sound waves?

6.  Briefly explain the benefits of wearing polarized sunglasses.

# THERMODYNAMICS

## WHAT YOU WILL LEARN

- The difference between heat and temperature.

- What specific and latent heats are.

- How the ideal gas law links the microscopic and the macroscopic.

- How to calculate the work done by a gas.

- How entropy relates to time.

- The four steps in a heat engine cycle.

| LESSONS IN CHAPTER 13 | |
|---|---|
| • Temperature and Heat | • Laws of Thermodynamics |
| • Gas Laws | • Engines |

## 13.1 Temperature and Heat

Temperature and heat are closely related, but they are not identical.

## WHAT'S THE DIFFERENCE?

**Temperature** is a measure of the average kinetic energy of the molecules in the substance being measured. Temperature is measured in degrees. Note, however, that although temperature is a measure of energy, degrees are not considered a unit of energy.

Although there are several different types of temperature scales, the proper SI units are for the absolute temperature, measured in Kelvin (K).

Kelvin is known as the **absolute temperature** scale because all motion ceases at zero Kelvin. Consequently, it is impossible to obtain a lower temperature, and zero Kelvin is known as **absolute zero**. Thus the Kelvin scale is correlated with the source of temperature: molecular motion. Although the Celsius and Kelvin temperature scales are different in so far as they have different zero points, the actual degrees themselves have the same magnitude:

$$Kelvin = Degrees\ Celsius\ (or\ Centigrade) + 273.15$$

Since the *change* in temperature in either the Kelvin or Celsius scale is the same, in any formula dependent on only the change in temperature, degrees Celsius may also be used.

In case you live in one of the few countries left still using the Fahrenheit scale, you must convert to Celsius or Kelvin before using any scientific equations. The magnitude of 1 degree is not the same in the Fahrenheit scale:

$$°C = \frac{5}{9}(°F - 32)$$

**Heat** is a measure of the energy being transferred from one substance to another because of a difference in temperature between the two substances. Heat, like other measures of energy, is typically measured in units of joules, although other units also exist, such as kilowatt-hours, BTUs (British thermal units), calories, etc.

Heat always flows from a hotter source to a cooler recipient. There are three forms of heat transfer.

1. Conduction: Heat is transferred by direct contact. The way in which heat flows through a material is determined by the material's **thermal conductivity**.

2. Convection: Heat is transferred by the actual movement of matter.

3. Radiation: Heat is transferred by electromagnetic waves being emitted and absorbed. The Stefan-Boltzmann equation shows that heat is radiated in proportion to the temperature raised to the fourth power. Dark surfaces have **high emissivity**, meaning they are both good emitters and good absorbers. Light or shiny surfaces, however, have **low emissivity**, meaning they are bad absorbers and bad emitters.

Consider this example: The sun heats Earth's surface by radiation. Then the surface of Earth heats the air in contact with it by conduction. Finally, the heated air rises and carries the heat to other areas via convection currents.

# HOW ARE HEAT AND TEMPERATURE RELATED?

Will heating a substance always produce a change in temperature? The answer is no. Heating produces a change in temperature only if the heat energy is absorbed as kinetic energy. In that situation, the substance's molecules move faster and the temperature increases. The **heat-temperature rule** governs this relationship:

$$Q = mc\Delta T$$

In the equation above:

- $Q$ = the heat being transferred (J)

- $m$ = the mass of the sample (kg)

- $c$ = the **specific heat** of the substance $\left( \dfrac{J}{kg \cdot degree} \right)$; an experimentally determined number that depends on the exact molecular nature of the substance

- $\Delta T$ = the temperature change (Celsius or Kelvin, since the change is the same)

If, however, the heat is absorbed as potential energy, the temperature of the substance does not change. This potential energy is usually associated with the molecules' electric attraction for each other. Heat that is absorbed in this nontemperature-changing manner is called **latent heat**.

# LATENT HEAT

Latent heat is associated with changes of state, or phase, in which the material is changing its structure. This latent heat energy is not apparent in the temperature; all the energy is being used to pull the molecules apart by overcoming the attractive forces. The rule for latent heat transfers is a simple one:

$$Q = mL$$

In the equation above, $L$ is the latent heat per unit mass of the specific material undergoing a change of state (J/kg). It is an experimentally determined number that depends on the exact molecular nature of the substance.

Figure 13.1 shows how the states of matter, temperature, and heat are related.

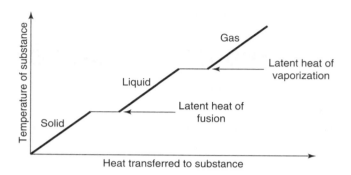

**FIGURE 13.1** The relationship between the states of matter, temperature, and heat.

The changes between states are defined as shown in Table 13.1.

**TABLE 13.1 CHANGES OF STATE**

| Solid | → | Liquid | | | Melting |
|-------|---|--------|---|---|---------|
| Solid | ← | Liquid | | | Freezing |
| | | Liquid | → | Gas | Evaporation |
| | | Liquid | ← | Gas | Condensation |
| Solid | | → | | Gas | Sublimation |
| Solid | | ← | | Gas | Deposition |

An **adiabatic** process is one in which no heat is gained or lost. For many heat transfer situations, one can make the approximation that the entire system of hotter and cooler objects is thermally insulated from the outside environment and thus adiabatic. In this case, heat will continue to flow from hot to cold until the objects come to **thermal equilibrium**, i.e., they share the same temperature. This process, when used to measure specific heats, is called **calorimetry**. The rule for the entire process is a simple one:

Heat Lost by Hot Object(s) = Heat Gained by Cool Object(s)

**EXAMPLE** How many degrees Celsius does a hot cup of tea ($m = 250$ g) lose as it melts 20 grams of an ice cube and comes to an equilibrium temperature? (Use the specific heat of water. The starting temperature of the tea = 85 degrees Celsius. The starting temperature of the ice cube = 0 degrees Celsius. The heat of fusion for ice = 335 J/g. The specific heat of water = 4.18 J/g • K.)

**SOLUTION**

Let the final equilibrium temperature be $x$:

$$\text{heat lost by tea} = \text{heat gained by ice cube}$$

$$(mC\Delta T)_{\text{tea}} = (mL)_{\text{ice}} + (mC\Delta T)_{\text{water from ice}}$$

$$250(4.18)(85 - x) = (20)(335) + (20)(4.18)(x - 0)$$

$$88,825 - 1,045x = 6,700 + 83.6x$$

Solve for $x$:

$$x = 73$$

Temperature loss for tea:

$$85 - 73 = 12 \text{ degrees}$$

# KINETIC ENERGY

The **kinetic theory** states that all molecules are in constant motion. The most famous evidence for this theory is called **Brownian motion** (1827), in which tiny particles are seen to be constantly in motion even when suspended in an apparently motionless liquid.

In 1905, Einstein studied Brownian motion. By using statistics, he calculated the size of atoms to be $10^{-10}$ m, which is the modern accepted value!

For an idealized simple gas, the following equation holds for each molecule:

$$\text{Average kinetic energy} = \frac{1}{2}mv_{\text{avg}}^2 = \frac{3}{2}kT$$

In the equation above:

- $v_{\text{avg}}^2$ = the average velocity squared of the molecules in the gas ($m^2/s^2$)

- $k$ = Boltzmann's constant ($1.38 \times 10^{-23}$ J/K)

- $T$ = the absolute temperature (K)

The square root of $v_{avg}^2$ is referred to as the **root-mean-square velocity** ($v_{rms}$), which is an indication of how fast the molecules are moving in the gas:

$$v_{rms} = \sqrt{v_{avg}^2}$$

**Boltzmann's constant**, $k$ or sometimes also referred to as $k_B$, provides the link from the microscopic details of a system to its macroscopic properties.

The kinetic theory explains why almost all materials undergo **thermal expansion** as they heat up. Because the molecules are moving faster, they collide more frequently with each other, thus taking up more space and causing the entire substance to expand.

For complicated molecules that can rotate or stretch, the above equation gives only the *translational kinetic energy*, which includes only back and forth, left and right, and up and down motion. Some energy will also go into the molecules' rotating and stretching!

Finding the **internal energy** or **thermal energy** (the sum total of all energy of all the molecules in an object) is simple in the case of the ideal monoatomic gas:

$$\text{Internal or thermal energy} = \frac{3}{2}NkT$$

In the equation above, $N$ is the number of molecules in the gas.

The internal energies of nonideal gases, liquids, and solids are much more complicated because you must include the electric interactions (potential energies) among the molecules.

# 13.2 Gas Laws

Recall the following:

- Pressure ($P$) $= \dfrac{\text{Force}}{\text{Area}}$

- Pressure is defined by units of **pascal** (Pa) equal to 1 $N/m^2$.

- Standard atmospheric pressure, 1 **atmosphere** (atm), is about $10^5$ Pa, which is also equivalent to 14.7 psi (pounds per square inch).

The pressure exerted by a liquid or gas on the sides of its container is due to the impact of the individual molecules. Early on, it was discovered that if you could determine the pressure, temperature, and volume of a gas, then the gas was completely described. To describe the gas, you must make the following assumptions.

1. There are a large number of molecules moving with random speeds in random directions.

2. Molecules are widely separated such that the only forces they exert on each other occur when they collide.

3. All collisions are perfectly elastic.

The complete description of an idealized gas is its **equation of state**:

$$PV = NkT$$

In the equation above, $V$ is the volume of the container holding the gas ($m^3$).

An **isothermal** process is one that takes place at constant temperature. The fact that pressure and volume are inversely proportional for a given gas if you hold the temperature constant is known as **Boyle's law**.

An **isochoric** or **isovolumetric** process is one that takes place at constant volume. The fact that pressure and temperature are directly proportional at constant volume is known as **Gay-Lussac's law**. There is a version of the heat-temperature rule that can be used for gases under isochoric conditions. In this case, one uses the molar heat capacity, $C_v$ (calories/degree/mole), rather than the specific heat, and so the rule becomes:

$$Q = nC_v\Delta T$$

In this equation, $n$ is the number of moles. (Moles are defined below.)

An **isobaric** process is one that takes place at constant pressure. The fact that volume and temperature are directly proportional for constant pressure is known as **Charles's law**. The comparable heat-temperature rule for isobaric processes is now the following:

$$Q = nC_p\Delta T$$

# CHEMISTRY JARGON

Since macroscopic materials are made of enormous numbers of molecules, chemists invented the concept of a mole. A **mole** is an enormous number of molecules. Specifically, it is the number that relates the **amu** (atomic mass units) of a single molecule or atom to grams. Recall that the amu of a molecule is found simply by counting the number of protons and neutrons in all of the atoms in the molecule. In other words, you add their atomic mass numbers (sometimes referred to as a molecule's formula mass or, even more disturbingly, formula weight). Since 1 amu $= 1.66 \times 10^{-24}$ g (about the mass of a proton), it takes $6.02 \times 10^{23}$ (**Avogadro's number**) molecules to make a mole. One molecule would have its mass measured in amu, whereas one mole of the same molecule would have the same number of grams as its mass. Moles are convenient to work with because grams of samples are easily measured in a lab, and one can determine the number of moles easily by dividing the sample mass by the molecule's molar mass.

Chemists prefer to call the following equation of state the **ideal gas law**:

$$PV = nRT$$

Note that the ideal gas law is very similar to the general equation of state. In the ideal gas law:

- $n =$ the number of moles

- $R =$ the universal gas constant ($8.315$ J/mole/K $= kN_A$)

Because of the ideal gas law, the two molar heat capacities mentioned above are related:

$$C_p = C_v + R$$

Table 13.2 defines the various thermodynamic processes.

### TABLE 13.2 THERMODYNAMIC PROCESSES

| Process | Definition |
| --- | --- |
| Isobaric | Constant pressure |
| Isochoric/isovolumetric | Constant volume |
| Isothermal | Constant temperature |
| Adiabatic | No heat transfer |

# 13.3 Laws of Thermodynamics

Various texts on physics will define and elaborate slightly different versions of thermodynamics. Commonly, a zeroeth law or a third law will be presented to complete the set of laws from a technical perspective. However, the two central laws are universally presented as fundamental to the study of physics and science in general.

## FIRST LAW OF THERMODYNAMICS

Energy is conserved. More specifically, the change in internal energy ($\Delta U$) of a closed system is equal to the heat added ($Q$) to the system minus the work ($W$) done by the system:

$$\Delta U = Q - W$$

This is the generalization of the conservation of mechanical energy we studied back in Chapter 6. If an ideal gas is compressed or expanded while it is in thermal equilibrium with its surroundings, the work done by or to the ideal gas is the area under the curve in a pressure ($P$) vs. volume ($V$) diagram, as shown in Figure 13.2.

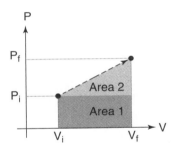

FIGURE 13.2 Pressure vs. volume diagram of an ideal gas.

Total work done by the gas to its surroundings is the total area under the graph.

**EXAMPLE** How much heat is generated when an ideal monoatomic gas is compressed from 0.75 m³ to 0.50 m³ and its temperature increases from 300 Kelvins to 320 Kelvins? (The gas is initially at atmospheric pressure.)

**SOLUTION**

Heat can be found once we calculate the change in internal energy due to the temperature change and the work done due to the compression:

$$\Delta U = Q - W$$

The work done *by* the gas will be negative since it is being compressed. We must find the corresponding pressure difference from the ideal gas law:

$$P_1 V_1 = NkT_1 \text{ and } P_2 V_2 = NkT_2$$

Since $N$ and $k$ are constant here, we can set the combination $\dfrac{P_1 V_1}{T_1}$ equal to $\dfrac{P_2 V_2}{T_2}$. (Note that we could also find $N$ from the first equation and use that value in the second: $N = \dfrac{P_1 V_1}{kT_1}$).

$$\frac{P_1 V_1}{T_1} = \frac{P_2 V_2}{T_2}$$

$$P_2 = P_1 \left( \frac{V_1}{V_2} \right) \left( \frac{T_2}{T_1} \right)$$

Plug in our values:

$$P_2 = 10^5 \left( \frac{0.75}{0.50} \right) \left( \frac{320}{300} \right) = 1.6 \times 10^5 \text{ Pa}$$

The work done is the area under the pressure vs. volume graph:

$$\text{Area 1} = 10^5 (0.75 - 0.50) = 2.5 \times 10^4 \text{ J}$$

$$\text{Area 2} = \frac{1}{2}(0.75 - 0.50)(1.6 - 1)\left(10^5\right) = 7.5 \times 10^4 \text{ J}$$

Total work done to the gas is $1 \times 10^5$ J.

Total work done by the gas is $-1 \times 10^5$ J.

Use the formula for the internal energy of a monoatomic gas and substitute the expression for the number of atoms from the ideal gas law:

$$\Delta U = \frac{3}{2} Nk \left( T_2 - T_1 \right) = \frac{3}{2} \left( \frac{P_1 V_1}{kT_1} \right) k \left( T_2 - T_1 \right) = \frac{3}{2} \left( \frac{P_1 V_1}{T_1} \right) \left( T_2 - T_1 \right)$$

$$\Delta U = \frac{3}{2} \left( 10^5 \right) (0.75) \left( \frac{320 - 300}{300} \right) = 7.5 \times 10^3 \text{ J}$$

The heat added to the gas is given by the following:

$$Q = \Delta U + W = (-1 \times 10^5 \text{ J}) + (7.5 \times 10^3 \text{ J}) = -0.925 \times 10^5 \text{ J}$$

The negative sign indicates that heat is leaving the gas. Note that a lot of work is done to the gas, but the internal energy does not rise much. So the excess energy leaves the system.

# SECOND LAW OF THERMODYNAMICS

The second law of thermodynamics is arguably one of the most powerful and yet subtle laws in all of science. As such, there are many formulations of it. At its heart is a statement about statistics and how most real-world processes are not completely reversible. As objects interact and exchange various forms of energy, there are many more ways to store the energy in unorganized, hard-to-recover forms than there are organized, easy-to-recover forms. Here are four different versions of the second law:

- Heat flows from hot to cold and will not spontaneously flow from cold to hot. (Clausius version)

- An engine cannot be constructed whose only effect is to transform a given amount of heat completely into work. (Kelvin-Planck version)

- Natural processes tend toward states of greater disorder. (General version)

- The total entropy of any isolated system increases during any natural process. (Entropy version)

**Entropy** is the measure of order or disorder of a system. Clausius determined that the change in entropy ($\Delta S$) is equal to the amount of heat added to it ($Q$) by a reversible process at a constant temperature $T$ (in Kelvin):

$$\Delta S = \frac{K}{T}$$

The entropy version of the second law of thermodynamics is said to be the only law of physics that actually defines a direction to time. Time going forward *is* universal entropy increasing! The reason we do not see certain processes that are allowed under other laws of physics is because the second law of thermodynamics would be violated. (Imagine two side-by-side broken pieces of a saucer spontaneously resealing.) Note that the measure of disorder might not correspond to our usual idea of disorder. What physicists mean by disorder is actually an increase is the number of microstates being occupied by all the particles in a system. A microstate is the actual position and velocity (or other key attributes that describe a particle) of the particle. The second law can be thought of as a spread in the possible configurations of equal energy, as shown in Figure 13.3.

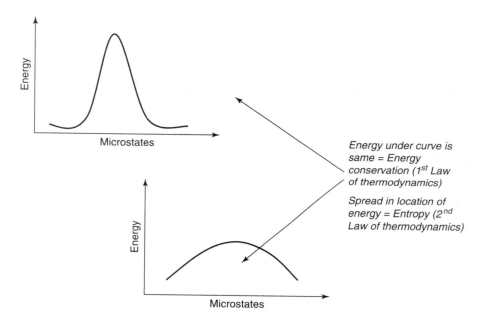

**Figure 13.3** Entropy and Energy.

As time moves forward, energies are being redistributed in more and more ways. As another way of thinking about it, organized energy is increasingly transformed into unorganized energy. For example, think of the overall, organized kinetic energy of a ball in motion compared with the unorganized kinetic energies of the ball's individual molecules in motion.

## THINGS TO THINK ABOUT

Here is a colloquial way of thinking about the first two laws of thermodynamics:

First Law: You can't get something for nothing.

Second Law: You won't even break even.

# 13.4 Engines

A **heat engine** is a machine that transforms some heat ($Q_H$) flowing from a higher temperature source ($T_H$) into work ($W$) plus some residual heat ($Q_L$) flowing to a lower-temperature ($T_L$) destination. The efficiency of a heat engine is defined as follows:

$$\text{Efficiency} = \frac{W}{Q_H}$$

The work done by a heat engine is equal to the following:

$$W = (Q_H - Q_L)$$

Therefore, efficiency can be described as shown here:

$$\text{Efficiency} = 1 - \frac{Q_L}{Q_H}$$

If the engine consists of completely reversible processes (which is impractical in reality because of friction), it is called a **Carnot engine** and is said to be ideal. The ideal efficiency of a Carnot engine can be calculated:

$$\text{Ideal efficiency} = 1 - \frac{T_L}{T_H}$$

The Carnot engine must cycle through four distinct steps:

1. Isothermal expansion

2. Adiabatic expansion

3. Isothermal compression

4. Adiabatic compression

Figure 13.4 shows these four distinct steps.

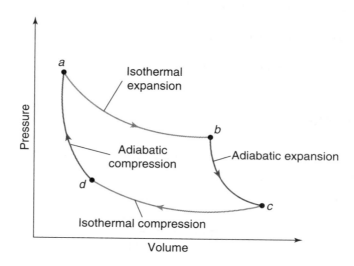

**Figure 13.4** The four distinct steps of a Carnot engine.

# Chapter Review Exercises

1. How many grams of water will be turned into steam if 7,500 calories are added to 500 grams of water at 90 degrees Celsius? (The heat of vaporization of water is 540 calories/gram.)

2. How many molecules are there in 3.0 liters of an ideal gas at 85 degrees Celsius if the absolute pressure is 2.5 atmospheres?

3. What is the average kinetic energy of the molecules of anything at room temperature? If the molecule is nitrogen gas, what is its root-mean-square velocity?

4. What equilibrium temperature will be reached if you pour 550 grams of hot water (97 degrees Celsius) into an isolated 750-gram aluminum pot at 22 degrees Celsius? (The specific heat of aluminum is 0.902 J/g • K.)

5. If you obtain 150 grams of ice from a freezer set to –9 degrees Celsius, how much heat does the ice absorb from the room while warming up to 21 degrees? (The specific heat of ice = 2.1 J/g • K; heat of fusion = 335 J/g; specific heat of water = 4.18 J/gK.)

6. At sea level, 220 liters of helium are put into a balloon at a temperature of 28 degrees Celsius. After the balloon rises to where the pressure has dropped to 35 kilopascals and the temperature is –11 degrees Celsius, what volume will the balloon now occupy?

7. A steam engine operates using steam at a temperature of 175 degrees Celsius, and its exhaust is sent into a river at 11 degrees Celsius. What would be the engine's ideal efficiency? Is it likely to reach this efficiency in actual operation?

8. Earth radiates longer-wavelength radiation back into space after absorbing slightly shorter wavelengths of radiation from the sun during the daytime. This phenomenon is an important piece of the global warming phenomenon. However, it is a great example of the second law of thermodynamics at work. Explain how this is an example of the universe's entropy increasing.

# FLUIDS

## WHAT YOU WILL LEARN

- What atmospheric pressure is and what causes it.

- What determines air friction.

- What terminal velocity is.

- How hydraulic systems work.

- What causes buoyancy and how to calculate it.

- What causes lift.

- Work done by a gas.

| LESSONS IN CHAPTER 14 | |
|---|---|
| • What Is a Fluid? | • Archimedes and Buoyancy |
| • Fluid Friction | • Bernoulli and Fluid Flow |
| • Pascal and Hydraulics | |

## 14.1  What Is a Fluid?

Generally speaking, a fluid is matter that takes the shape of its container. Rather than being primarily described by its mass, a fluid is usually categorized by pressure and density:

$$\text{Pressure} = \frac{\text{Surface Force}}{\text{Area}}$$

$$\text{Density} = \frac{\text{Mass}}{\text{Volume}}$$

The pressure of a fluid is a function of the weight of the fluid above a certain cross-sectional area, and thus the pressure increases as a function of the depth of fluid. Most liquids are thought of as incompressible fluids (meaning they maintain their density), whereas gases are categorized as compressible fluids (their density changes).

## THINGS TO THINK ABOUT

*Atmospheric pressure*

We all live at the bottom of an enormous fluid: the atmosphere. As such, we constantly experience a pressure exerted on us: atmospheric pressure. Baseline atmospheric pressure is taken to be the pressure Earth's atmosphere exerts at sea level. For a static fluid, the pressure is due to the weight of the column of fluid above it. Thus the atmospheric pressure is reduced as you go to higher and higher altitudes. Earth's atmospheric pressure (or 1 atmosphere) is about $10^5$ pascals. (In other words, 1 square meter at sea level has about $10^5$ newtons of atmosphere above it.) Other historical units for atmospheric pressure include 14.7 psi (pounds per square inch) and 760 torr (based on the millimeters that mercury will rise in a partially enclosed tube when exposed to atmospheric pressure).

EXAMPLE How deep under water must you dive to experience 1 additional atmosphere of pressure due to the water pressure?

### SOLUTION

The pressure is the weight of the fluid above per unit area:

$$\text{Weight} = mg = (\text{density})(\text{volume})g = (\text{density})(\text{unit area})(\text{height})g$$

$$\text{Pressure} = \frac{\text{weight}}{\text{unit area}} = \rho g h$$

In the second equation, $\rho$ is the density of the fluid and $h$ is the height of the fluid column above the unit area. In this problem, we use the density of water and standard air pressure:

$$\text{Pressure} = 1 \text{ atmosphere} = 10^5 \text{ pascals} = (1,000 \text{ kg/m}^3)(9.8 \text{ m/s}^2)h$$

Solve for $h$:

$$h = \frac{100}{9.8} = 10.2 \text{ m}$$

# 14.2 Fluid Friction

Friction exerted by fluids on objects traveling through them is a more complicated and dynamic force than the sliding friction experienced by two solid objects scraping their surfaces by each other that we studied in Chapter 2. This fluid friction force is known as drag, and it depends on three things: the viscosity of the fluid, the speed of the object's motion through the fluid, and the size and shape of the object. The **viscosity** of a fluid describes how easily the fluid itself flows. Clearly, a fluid with low viscosity will exert a smaller drag force on an object passing through it than a fluid with high viscosity. Additionally, as the fluid flows around the surface of the object traveling through it, the flow can be categorized as either laminar or turbulent. **Laminar** flow literally means the fluid is flowing in organized layers, and this type of flow produces lower drag forces. Laminar flow happens around objects that are categorized as streamlined. **Turbulent** flow, on the other hand, exerts larger drag forces as the fluid makes its way around the object in a more chaotic fashion.

Laminar flow, as shown in Figure 14.1, requires that the volume of fluid moving with the flow must be the same. In terms of the cross-sectional areas ($A$) to the flow, the velocity ($v$) of flow must change in order to maintain the same volumetric flow rate:

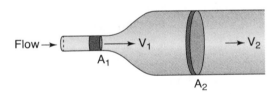

**FIGURE 14.1** Laminar flow.

Earlier, in our study of kinematics and simple forces, we modeled objects falling through the air in the absence of friction. However, the atmosphere is a fluid and does exert a drag force, which can be significant if you have ever put your hand out of the window in a speeding car. This air drag is the reason that objects falling through the air frequently do not fall at constant acceleration. Rather, they accelerate at an ever-decreasing rate as their speed through the air increases until they acquire **terminal velocity**. As the object speeds up, the fluid friction or air drag also increases, which decreases the net force. This can continue until the air drag completely cancels out the downward force of gravity, resulting in a net force (and acceleration) of zero. Since the object will no longer change its speed, it will fall at the same speed for the rest of its journey, hence the name for this speed—terminal velocity.

**EXAMPLE**    A 15 kg object reaches a terminal velocity of 8 m/s after being thrown off a tall cliff. What is the air drag acting on the falling object?

   **SOLUTION**

   The free-body diagram of a falling object experiencing air friction is shown here.

   Since the object is no longer accelerating, the net force must be zero:

$$F_f - F_g = ma = 0$$

$$F_f = F_g = mg = 15(9.8) = 147 \text{ N}$$

# 14.3 Pascal and Hydraulics

Pascal's principle is that any change in pressure at any location in a fluid will be transmitted uniformly to all parts of the same fluid. For instance, if a fluid is enclosed and an external force $F_1$ is applied to a certain area $A_1$ on the surface of the fluid, the resulting increase in force $(F_2)$ at a different location with its own cross-sectional area $A_2$ can easily be determined:

$$\frac{F_1}{A_1} = \text{change in pressure} = \frac{F_2}{A_2}$$

This is the principle underlying all hydraulic systems, such as brake lines in cars, wheelchair lifts, and forklifts. Because the forces and power available via fluid pressure are considerable, hydraulics are frequently used in industrial applications and tasks involving heavy equipment.

**EXAMPLE**    If the narrow end of hydraulic line is 0.05 m² and experiences a force of 6 N, what force will the wider end (0.25 m²) experience?

   **SOLUTION**

$$\frac{F_1}{A_1} = \frac{F_2}{A_2}$$

$$\frac{6}{0.05} = \frac{F_2}{0.25}$$

$$F_2 = 30 \text{ N}$$

# 14.4 Archimedes and Buoyancy

Buoyancy is the net upward force applied to a submerged (or floating) object in a fluid due to the difference between the higher pressure on the lower portion of the submerged object and the lower pressure on the higher portion of the submerged object. Archimedes determined that this buoyant force is equal to the weight of the fluid being displaced by the submerged portion of the object:

$$F_{\text{buoyant}} = (\text{volume of object in fluid})(\text{density of fluid})g = V_s \rho_f g$$

**EXAMPLE** A volleyball has a mass of 0.38 kg. As it floats in a pool of water, what volume of the ball lies beneath the surface?

**SOLUTION**

Since the ball is at equilibrium, the buoyant force upward of the water on the ball must be the same as the downward force due to gravity on the ball:

$$F_b = mg$$
$$V_s \rho_f g = mg$$
$$V_s = \frac{m}{\rho_f}$$

Use the approximate density of water, 997 kg/m$^3$:

$$V_s = \frac{0.38}{997} = 0.00038 \text{ m}^3$$

Since the volume of a volleyball is about 0.0046 m$^3$, less than 9% of the volleyball is beneath the waterline of the pool.

Note that the surface of a liquid acts like an elastic membrane under very mild forces. This surface tension is a different phenomenon from buoyancy. When a light object like an insect is supported by surface tension, it is a property of the surface of the liquid and is not related to pressure. Surface tension causes water to contract to minimize its surface area, forming spherical droplets. Detergents have the property of breaking this surface tension so that the water is allowed to penetrate into smaller areas than its surface tension will normally allow.

# 14.5 Bernoulli and Fluid Flow

As a fluid moves, the energies of the molecules change, which, in turn, affect the pressure the fluid exerts on its surrounding. By taking into account the kinetic and gravitational potential energies of the moving fluid, along with enforcing some continuity conditions on the fluid flow, Bernoulli came up with the following conserved quantity for the fluid flow:

$$P + \rho g h + \frac{1}{2}\rho v^2$$

In this equation, $P$ is the pressure, $h$ is the height, and $v$ is the velocity of the fluid.

If the fluid is not moving, the $v$ is zero. We recoup the simple explanation of the pressure increasing as a function of the weight of the fluid above that point. If we compare two points at the same height, we see that when the velocity of the fluid is bigger, the pressure is lower.

In this one equation, then, we have explanations for a variety of phenomena. Here are a few common examples with their explanation in terms of the Bernoulli equation.

**EXAMPLE 1** Partially covering the opening of a hose increases the velocity of the fluid but requires a great deal of force to maintain.

Let position 1 be before the constriction and position 2 be at the constriction:

$$P_1 + \rho g h_1 + \frac{1}{2}\rho v_1^2 = P_2 + \rho g h_2 + \frac{1}{2}\rho v_2^2$$

Since the stream of water is at the same height just before and after the partially closed off opening, $h_1 = h_2$ and the equation can be simplified:

$$P_1 + \frac{1}{2}\rho v_1^2 = P_2 + \frac{1}{2}\rho v_2^2$$

Since the flow is continuous and the cross-sectional area $(A_2)$ has been made smaller, the initial velocity $v_1$ must be larger than the final velocity $v_2$ by the same factor that the area has changed:

$$A_1 v_1 = A_2 v_2$$

Since this increase in velocity is squared in the Bernoulli equation, the initial pressure $P_1$ must be much higher than the final pressure $P_2$. The general lowering of pressure when the cross-sectional area gets smaller is called the **Venturi effect**.

This difference in pressure $(P_1 - P_2)$ is the source of the great force needed to restrict the cross-sectional area in our hose example.

**EXAMPLE 2**  Blowing across the top of a piece of paper can cause the paper to rise (or a flag to wave in the wind).

Once again, the heights are basically the same on either side of the paper (or flag):

$$P_1 + \frac{1}{2}\rho v_1^2 = P_2 + \frac{1}{2}\rho v_2^2$$

The side experiencing the high rate of fluid flow must have a reduction of pressure on that same side in order to maintain the equality. The difference between the high pressure on the opposite side and the lower pressure on the side with faster fluid flow exerts a force, causing the paper (or flag) to move. This difference in pressure between the bottom and top side of an airplane's wing is part of the source of the lift force that allows airplanes to fly.

**EXAMPLE 3**  Poking a hole at the bottom of a can of liquid will cause the liquid to eject at high velocity:

$$P_1 + \rho g h_1 + \frac{1}{2}\rho v_1^2 = P_2 + \rho g h_2 + \frac{1}{2}\rho v_2^2$$

Approximate that the liquid at the top of the can is not moving significantly, so $v_1 = 0$. Take the height to be zero at the level of the hole: $h_2 = 0$.

If the can is open at the top, the pressure at both the top and the hole is the same (atmospheric pressure: $P_1 = P_2$):

$$\rho g h_1 = \frac{1}{2}\rho v_2^2$$

Solve for $v$ at the hole in terms of the depth of the hole in the liquid ($h$):

$$v = \sqrt{2gh}$$

During a sudden gust of strong wind, the fire in your fireplace gets brighter and taller.

First note that even without any difference in flow rate (if $v_1 = v_2$), just the fact that the top of the chimney is at a bigger $h_2$ value implies that the pressure at the top of the chimney is lower:

$$P_1 + \rho g h_1 + \frac{1}{2}\rho v_1{}^2 = P_2 + \rho g h_2 + \frac{1}{2}\rho v_2{}^2$$

This explains why there is almost always a draft up a chimney.

In the case of a strong wind outside, the fluid flow is much higher at the top of the chimney ($v_2 > v_1$). As $v_2$ increases, $P_2$ must decrease in order to maintain the equality. Therefore, the pressure difference between the higher-pressure bottom of the chimney and the lower-pressure top increases, which pushes more air through the chimney. This, in turn, increases the oxygen fanning the flames, causing them to burn more fiercely.

# Chapter Review Exercises

1. Calculate the difference in air pressure between the bottom and top of a vertical 20-meter pipe. The density of air is around $1.3 \text{ kg}/\text{m}^3$.

2. If a horizontal pipe has water flowing through it and is gradually constricted to half its cross-sectional area, by what factor does the pressure in the pipe change?

3. What is the upward normal force on a 5.00 kg brick (dimensions 215 mm $\times$ 102.5 mm $\times$ 65 mm) resting at the bottom of a shallow pool?

4. Explain why a fast-moving train that passes close by tends to draft things toward it.

5. Explain how a ship made of metal can float in water even though metal is denser than water.

6. Explain why a gust of wind over you may cause your umbrella to fold upward.

7. What percentage of an ice cube floats above the surface of the water if it is floating in water? ($900 \text{ kg}/\text{m}^3$ is the density of ice.)

# QUANTUM PHYSICS

## WHAT YOU WILL LEARN

- What everything is made of.
- Planck's solution to the ultraviolet catastrophe.
- Einstein's solution to the photoelectric effect.
- History of wave-particle duality.
- What Schrodinger's waves, Heisenberg's uncertainty, and Feynman's paths have in common.
- Radioactivity and half-lives.

| LESSONS IN CHAPTER 15 | |
|---|---|
| • The Standard Model | • Nuclear Physics |
| • Early Quantum Mechanics | • Classical vs. Modern Physics |
| • Complete Quantum Theories of the Atom | |

## 15.1 The Standard Model

What are things made of? That is probably the most basic of all science questions. We now know the answer: quantum fields. This unexpected answer is neither intuitive nor easy to explain. However, careful experimentation and heavy-duty theoretical lifting on the part of physicists worldwide over the past 120 years have led us to this conclusion.

For every step forward, there seems to be a temporary increase in complexity, which then leads to another insightful break that ultimately simplifies our understanding of how everything ties together.

Consider the excitement of early chemists discovering that everything is made of just a limited number of elements. This insight was then followed by an explosion of types of elements collected elegantly into the periodic table as well as a staggering number of basic parts out of which to create all matter. However, the plethora of elements reduced itself eventually to the insight that all elements are just an accumulation of protons, neutrons, and electrons. Then, for a brief time, it seemed all matter could be explained with just those three subatomic particles.

Naturally the question then arises, are these three subatomic particles the answer to the question of what matter is made of? The story does not end with the idea that all matter is made of elements, which are in turn made of protons, neutrons, and electrons, By slamming these subatomic particles into each other and carefully observing the particles produced during radioactive decay, we discovered that although electrons do not have any substructure and are indivisible, protons and neutrons are composite particles. The component parts of protons and neutrons are fundamental particles called quarks. Painstakingly, over the past 100 years, a list of fundamental particles that is thought to be complete has been assembled into what is known as the standard model. Figure 15.1 is a picture of a subset of the standard model, which includes all the known building blocks of stable matter in the universe. Note there are two higher generations of fermions, but they are unstable and quickly decay into one of the first-generation particles pictured in Figure 15.1.

These **fundamental particles** are the end of the line so far as we know. They do not have substructures or components within them. If we try to break them into smaller pieces, they simply annihilate and turn into energy. The list also appears to be complete within its domain: all observed matter and interactions involving nuclear and/or electromagnetism are made of some combination of these particles. Note that a quantum theory of gravity remains elusive, and Einstein's general theory of relativity (which is not quantum) reigns supreme in explaining gravitational interactions. However, there are gravitationally related mysteries such as dark matter and dark energy that hint at a more complete theory of gravity yet to come. See Chapter 16 for more about these.

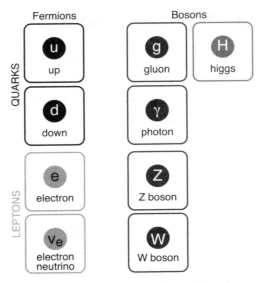

**Figure 15.1** Simplified Standard model of elementary particles.

There are many ways in which to divide and organize this list of fundamental particles. However, as can be seen in Figure 15.1, they all fall into one of two categories: fermions or bosons.

**Fermions** make up the material that we traditionally called matter. Fermions all obey the Pauli exclusion principle. Most students first encounter this concept when studying the first fundamental particle discovered: the electron. The Pauli exclusion principle states that fermions do not share quantum states. Therefore, they take up space when they are drawn to the same location.

**Bosons** are the stuff of forces. All the forces except for gravity can be explained by the five bosons in the standard model. (If a graviton is discovered, then gravity will also be included.) Bosons share quantum states and thus do not take up space.

# 15.2 Early Quantum Mechanics

The story begins with Max Planck taming what was known as the **ultraviolet catastrophe** in 1900. The problem was that when applying classical ideas of energy, the frequencies of light emitted by virtue of an object's temperature (known as its **blackbody radiation**) would increase as the wavelengths of radiation got shorter. Recall that accelerating charges release electromagnetic radiation. Thus all atoms in vibration due to their thermal motion must be emitting radiation. Although experimentally it was shown that all objects emitting electromagnetic radiation were emitting various wavelengths of electromagnetic radiation as a function of their temperature, real objects were emitting far less low-wavelength (ultraviolet) light than was predicted by the physics of the day. In a fit of mathematical inspiration, Planck quantized the allowed energies of vibration. This quantization of energy resolved the issue:

$$E = nhf$$

In this equation, $h$ became known as Planck's constant, which links the frequency of oscillation to the energy emitted. The number of quanta available, $n$, is an integer. This idea of quantities in nature coming in discrete chunks rather than being continuous is precisely what is meant by **quantization**. Charge had just been shown to be quantized with the discovery of the electron a few years earlier, and so the quantum revolution had begun. For his contribution of the idea of energy quantization, Planck was awarded the Nobel Prize in Physics in 1918.

In 1905, Einstein took Planck's mathematical trick and turned it on its head by proposing that the quanta represented a physical particulate nature to light and to electromagnetic radiation in general. This reversal of the current understanding of light as a wave enabled Einstein to explain the **photoelectric effect** in which certain photosensitive metals eject electrons in numbers (but not energies) proportional to the intensity of light rather than the frequency of light. The frequency of light was responsible for the existence of a cut-off frequency and dictated the energy of each individual electron ejected. By quantizing light into what we now call photons, Einstein explained the energy of the ejected electrons by proposing a one-to-one relationship between quanta of light and quanta of charge:

$$\text{Kinetic energy of ejected electrons} = hf - W$$

In this equation, the $hf$ from Planck's proposal is now interpreted as the energy of a single photon and $W$ is the binding energy of the electron in the metal (the work required to free the electron from its metal). Higher intensity of light (without changing the color or frequency) results in more one-to-one interaction and thus more electrons will be ejected. However, each ejected electron will have the same energy as expressed in Einstein's formula. For this work, Einstein was awarded the Nobel Prize in Physics in 1921.

In 1923, to bolster Einstein's photon (particle-like) interpretation of light, Arthur Compton supplied further experimental proof of the particle-like nature of light. He showed that photons have momentum (like particles) by observing conservation of momentum in a collision between photons and electrons. Like Einstein's formulation of the energy per photon, Compton showed that each photon has a momentum:

$$\text{momentum of photon} = p = \frac{h}{\lambda}$$

Thus Planck's constant began to emerge as a fundamental constant of nature, alongside of the speed of light ($c$) and the universal gravitational constant ($G$), by describing how the quanta are sized. Compton received the Nobel Prize in Physics in 1927 for this discovery.

Inspired by the ubiquity of symmetry in nature, Louis de Broglie proposed that if light (thought of as wave-like) had particle-like properties, then matter should demonstrate wavelike properties. Also in 1923, he proposed that some of the strange behaviors of electrons inside the atom could be explained by giving them wavelike properties. Specifically, de Broglie accomplished this by reversing Compton's momentum formula and assigning all particles a wavelength:

$$\text{Wavelength of particle} = \lambda = \frac{h}{\text{momentum}} = \frac{h}{p}$$

For his proposal of this idea and his theoretical work showing that the stable orbits of an electron could be interpreted as standing waves of electrons around the nucleus, de Broglie was awarded the Nobel Prize in Physics in 1929. This inspired some of the most famous experiments in quantum mechanics, such as the diffraction patterns that can be observed when firing only one electron at a time at a double-slit setup.

The storied history of the battle in physics between interpreting light as a wave or a particle thus ended in the most unforeseeable synthesis: all things have both wavelike and particle-like properties. Niels Bohr, widely regarded as the philosopher-in-chief of the early years of quantum mechanics, weighed in with this deep observation about the seemingly bizarre synthesis, "There exist two sorts of truth: trivialities, where opposites are obviously absurd, and profound truths, recognized by the fact that the opposite is also a profound truth." Bohr is credited with the idea of **complementarity** in quantum mechanics. He said that wave nature and particle nature are two sides of the same coin and that an object will display the properties the experimenter is looking for. We cannot hold both visualizations at the same time, so choose the one that is appropriate to the phenomenon you are observing.

Although the modern view is that all quantum objects can be described as waves or particles, whether to treat light as a series of particles or as a continuous wave was a raging debate in the 1800s and 1900s. When Newton speculated on the nature of light, he decidedly supported the particle theory in opposition to Christiaan Huygens, who thought all light phenomena should be explained by the propagation of waves. Thomas Young's double-slit experiment (described in Chapter 12), however, showed (decisively it was thought, until Einstein) that light was a wave by demonstrating interference. By combining this evidence with the elegance of Maxwell's equations, which demonstrate convincingly that light is a transverse wave of electric and magnetic energies, light's ultimate nature as a wave seemed secure until Einstein's explanation of the photoelectric effect in 1905. The timeline of evolution of **wave-particle duality** (see Figure 15.2) is instructive in that it shows that science does not always progress by overturning previous theories; sometimes progress is made by realizing both viewpoints are valid!

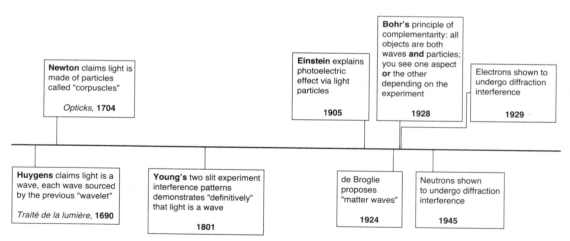

**Figure 15.2** Waves vs. particles timeline.

With our modern understanding that mass and energy are exchangeable (via $E = mc^2$) along with the quantum world's demonstration that waves and particles are somehow exchangeable as well, photons of electromagnetic radiation can generate matter spontaneously and vice versa as long as relativistic energy, charge, momentum, and all the other conservation laws are obeyed. In order for nature to accommodate the conservation laws, these types of exchanges necessitate the existence of antimatter. **Antimatter** exists and has the same properties as regular matter only with opposite charge and certain other properties. Paul Dirac predicted the positron (the positively charged antimatter version of an electron) in 1931; Carl Anderson proved it existed in 1932. Since that time, the antimatter counterpart of every particle in the standard model has been demonstrated to exist. The modern question to ask is not why does antimatter exist but, rather, why do we live in a universe dominated by what we consider "regular" matter such that when antimatter is created, it quickly encounters a matter version of itself and they annihilate, creating photons of equivalent energy? The source of this imbalance is an open question in physics.

Alongside of the development of wave-particle duality, the early 20th century was also a time of applying quantum ideas to the structure of the atom itself. Spectroscopy, which is the study of light interacting with atoms, was showing that electrons were confined to certain discrete energy levels within the atom. Bohr expanded the uses of quantum ideas by matching the spectroscopy of the hydrogen atom by quantizing the angular momentum:

$$\text{Angular momentum of the bound electron} = L = \frac{nh}{4\pi}$$

For this and his other contribution to our understanding of the atom, Bohr was awarded the Nobel Prize in Physics in 1922. Note that the image often presented of an electron orbiting around the nucleus in its allowed orbit like a planet orbiting the sun is deeply problematic. Planetary orbital motion is circular motion that requires acceleration. Accelerating charges produce electromagnetic radiation that carries away energy. Thus, if electrons were actually orbiting inside an atom in such a conventional way, they would lose energy continuously and experience orbital decay. Apparently, the quantum world is even stranger than simply just chopping up quantities previously thought of as continuous into discrete chunks.

# 15.3 Complete Quantum Theories of the Atom

After the early years of quantum mechanics demonstrated the need for a completely different model in science than the classical models, three different but equivalent theories of quantum mechanics emerged. Each is predicated on different principles but is complete in its own way. All three are beyond the scope of an introductory physics class, but a brief sketch of each is presented here.

# HEISENBERG AND THE MATRIX-DRIVEN MATHEMATICAL MODEL

Heisenberg and his model are most famous for demonstrating a certain granularity in the possibility of knowing or measuring any pair of conjugate variables (see Noether's theorem Chapter 6). This fundamental indeterminacy is known as the famous Heisenberg uncertainty principle. The following inequality shows the most common conjugate pair (position, $x$, and momentum, $p$):

$$\Delta x \Delta p \geq \frac{h}{4\pi}$$

Although it is possible to determine or know the position or momentum of a particle with arbitrary accuracy, once measurements are down at the quantum level (as defined by Planck's constant in the above relationship), this knowledge comes at the cost of increasing the indeterminacy of the other variable. Although sometimes presented as an experimental limitation, this is in actual fact a limiting feature of the quantum world a priori and naturally follows once a wavelike component to describing nature is adopted.

# SCHRODINGER AND THE WAVE EQUATION

Schrodinger is most famous for coming up with a continuous equation where the solution is the probability amplitude for finding an electron at a given location. The solutions to his wave equation are three-dimensional volumes of probabilities of various shapes that are known as the orbitals of electron location inside of the atom. The shape and periodicity of the table of elements with the specific nature of $s$, $p$, $d$, and $f$ orbitals can be derived from the Schrodinger wave equation.

# FEYNMAN PATH INTEGRALS

Feynman expanded the classical concept of the path of least action. Instead of the classically deterministic path whereby the unique path of the particle is determined, each possible path in the quantum world is followed in principle, but each is weighted by its own unique probability.

# SUMMARY OF THE THREE MODELS

Although all three formulations of quantum mechanics are mathematically and philosophically different, they have been shown to be equivalent. What they share is a fundamentally indeterministic or probabilistic interpretation of the quantum world. Although the philosophical implications and possible interpretations of these features of nature are not agreed upon by physicists, the accuracy of these models in predicting the results of experiments is unparalleled in science. So regardless of how unsettling we may find this baked-in feature of indeterminacy, numerous experiments have demonstrated it to be true.

# 15.4 Nuclear Physics

The nuclei of atoms are a melting pot of quarks that make up the protons and neutrons and of the gluons that constitute the strong nuclear force that binds quarks together. A proton is made of two up quarks and one down quark. A neutron is made of two down quarks and one up quark. From the outside of the nucleus, one can track the changes that happen during nuclear reactions by simply counting the neutrons and protons. The standard notations are as follows:

$$Z = \text{atomic number} = \text{number of protons in nucleus}$$

$$A = \text{mass number} = \text{number of nucleons (protons plus neutrons) in nucleus}$$

$$^{A}_{Z}\text{Element symbol}$$

or

$$[\text{Element symbol}]\text{-}A$$

## THINGS TO THINK ABOUT

All objects emit radiation of some kind (see blackbody radiation above). However, if the radiation is considered dangerous (either because it is a photon of high enough energy to ionize or because it is a charged particle), a substance is considered radioactive. Radiation refers to the particles coming out of the radioactive source, and it is this radiation that has the potential to damage living organisms at a molecular level. The three common types of radiation are alpha (helium nuclei, highly ionizing but very easily stopped), beta (electrons or positrons, ionizing but blocked relatively easily), and gamma (high-energy photons, least ionizing but hardest to block).

If there exists a lower-energy state for the system to fall into, a nucleus will emit some kind of radiation as it passes from its higher-energy, unstable state to its lower-energy, more stable state. This process is called **radioactive decay**. The radiation emitted is determined by the conservation laws. As is everything in the quantum world, the "decay" of the nucleus is probabilistic rather than deterministic. As such, no one can predict when any one particular nucleus will undergo this decay. However, large sample sizes of unstable particles or unstable nuclei can be categorized by their half-life $\left(t_{\frac{1}{2}}\right)$. The **half-life** is the time it takes, on average, for half of the amount of material present to undergo its radioactive decay. When the rate at which something occurs is proportional to the amount of substance, that rate is characterized by an exponential function:

$$N(t) = N_0 e^{-t/\tau}$$

In this equation, $N(t)$ is the number of unstable particles that remain from the original amount $N_0$ after time $t$ has passed. The variable $\tau$ is the mean lifetime of the unstable particle and is related to the half-life as shown below:

$$t_{\frac{1}{2}} = \tau \ln(2)$$

EXAMPLE A free neutron has a mean lifetime of about 880 seconds before it decays into a proton. This process is called beta decay because an electron (sometimes called a beta ($\beta$) particle) is emitted to conserve charge. (An antineutrino is also emitted but is omitted here for simplicity.)

$$n \rightarrow p^+ + e^-$$

If a starting population of 1 million free neutrons is monitored for 20 minutes, statistically how many free neutrons would be expected to remain at that time?

**SOLUTION**

First determine the mean lifetime:

$$\tau = \frac{t_{\frac{1}{2}}}{\ln(2)} = 1{,}270 \text{ seconds}$$

Now plug all the givens into the exponential decay equation:

$$N(t) = N_0 e^{-t/\tau}$$

$$N(t) = 10^6 e^{-(20 \times 60)/1{,}270}$$

$$N(t) = 10^6 (0.389)$$

$$N(t) = 389{,}000 \text{ neutrons left}$$

# 1. BETA DECAY

In beta decay, $e$ a beta particle (an electron) is emitted from the nucleus as a neutron decays into a proton. This has the effect of increasing the atomic number while the mass number remains the same. For example:

$$^{32}_{15}\text{P} \rightarrow ^{32}_{16}\text{S}^+ + e-$$

Also considered to be beta decay is when a proton transforms into a neutron by emitting a positron (the antimatter version of an electron). This is also known as positron emission (and a neutrino is emitted as well):

$$p^+ \rightarrow n + e^+$$

# 2. ALPHA DECAY

The alpha particle is a helium nucleus (2 protons and 2 neutrons: He-4, $^4_2\text{He}$, or $^4_2\alpha$). When an alpha particle is emitted, the atomic number decreases by 2 and the mass number decreases by 4. Although high-mass nuclei can shed single nucleons, He-4 is the preferred means by which a nucleus sheds positive charges. This occurs because the binding energies are so strong within He-4 that it requires very little energy from the nucleus and is a very stable subunit within a large nucleus. Also, the radioactive elements that have longer half-lives tend to be very neutron rich, increasing the odds of an alpha particle being the unit of emission from radioactive material we are likely to encounter. For example:

$$^{210}_{84}\text{Po} \rightarrow ^{206}_{82}\text{Pb}^{2-} + ^4_2\text{He}^{2+}$$

# 3. GAMMA DECAY

In gamma decay, a high-energy photon is emitted from a high-energy nucleus. As the photon does not carry any charge or other nucleon attribute, the atomic number and mass number remain the same. The high-energy nucleus is made more stable by shedding its excess energy in this manner. Generally speaking, this type of decay happens very quickly once the high-energy nucleus is formed through some other means.

# 4. FUSION

Fusion is the process in which two lighter nuclei come together and join to create one larger nucleus. Lighter elements (generally speaking up to iron, element 26) have smaller nuclei. Thus the short-range strong nuclear force can create so much negative binding energy that the overall energy goes down as these lighter elements fuse into a heavier one. Most dramatic of these fusion events is the chain of nuclear reactions creating the energy emitted from the sun.

In the first step, single protons (hydrogen ions) in the plasma core of the sun fuse together to create deuterium (a heavier isotope of hydrogen). In addition, a positron and neutrino are emitted. The positron likely meets an electron shortly. They annihilate and convert their mass into energy, while the chargeless, super-low-mass neutrino flies directly out of the sun without any interactions:

$$\ce{^1_1H^+} + \ce{^1_1H^+} \rightarrow \ce{^2_1H^+} + e^+$$

In the second step, one of the deuterium ions fuses with another proton. They form an unstable isotope of helium and a gamma photon:

$$\ce{^1_1H^+} + \ce{^2_1H^+} \rightarrow \ce{^3_2He^{2+}}$$

Once the density of He-3 ions is high enough, even more energy is released by a third (and final, at least within our sun currently) nuclear reaction:

$$\ce{^3_2He^{2+}} + \ce{^3_2He^{2+}} \rightarrow \ce{^4_2He^{2+}} + 2\ce{^1_1H^+}$$

Other fusion reactions occur in hotter stars (see Chapter 16). In this way, all the heavier elements are created via fusion in our universe.

## 5. FISSION

Fission is the process in which a single, heavier nucleus breaks into smaller, lower-mass nuclei. Although the strong nuclear force is strong enough to overcome the electrostatic repulsion between protons, it is an extremely short-range force. If the nucleus is very large and contains many protons, it is much more likely to be unstable than stable. Energy is released as the nucleus of these higher-mass, unstable nuclei break into smaller "daughter nuclei."

The decay of uranium-235 is one of the more easily controlled fission reactions as it allows for a chain reaction in that more neutrons are released to trigger additional U-235 fission events:

$$\ce{^{235}_{92}U} + \ce{^1_0}n \rightarrow \ce{^{89}_{36}Kr} + \ce{^{144}_{56}Ba} + 3\ce{^1_0}n$$

# 15.5 Classical vs. Modern Physics

When including the contributions to physics of special relativity, general relativity, quantum mechanics, and some astrophysics, the modern view of the universe and how it operates is almost completely different from the classical one. This is shown in Table 15.1.

**TABLE 15.1 A COMPARISON OF CLASSICAL AND MODERN PHYSICS**

| Classical (Pre-20th Century) | Modern (Emerges Mid-20th Century) |
|---|---|
| Space and time exist separately and are universal and fixed in nature. | Space and time are two sides of the same coin, and they are aspects of the gravitational field. They bend and change, and they are observer dependent. |
| Everything is continuous. Space and time extend to infinity. | Everything is quantized. Time had a beginning; accessible space is finite. |
| Stuff is made of particles that have well-defined locations. Waves are ripples traveling through the stuff. | Everything has both wavelike and particle-like attributes. Fundamental particles do not have definite volumes, precise locations, or exact speeds. |
| Mass is conserved. | Total energy (including $mc^2$) is conserved, and mass is just one type of energy. Fundamental particles can be destroyed or created as long as the conservation laws are honored. |
| The universe is fixed and eternal. | The universe is 13.8 billion years old and is expanding at an ever-increasing rate. Stars are born and die (see Chapter 16). |
| Particles interact via forces. | All the players (particles and force carriers) are fundamental particles in the standard model, and they must obey the six conservation laws. All other possibilities can and do happen: <br><br> • Conservation of energy <br> • Conservation of linear momentum <br> • Conservation of angular momentum <br> • Conservation of charge <br> • Conservation of baryon number <br> • Conservation of lepton number |

# Chapter Review Exercises

1. Once Niels Bohr quantized angular momentum, his model of hydrogen's energy levels for electrons took the following form:

$$E_n = \frac{E_1}{n^2}$$

In this equation, $E_1$ is the energy of the lowest energy level (ground state), and $n$ is an integer representing the energy level.

If $E_1 = -13.6$ eV, determine the frequency of light emitted when an electron falls from $n = 3$ to $n = 1$ in this model. (Recall that eV is a unit of energy from electrical potential, $qV$).

2. Refer back to the section on Noether's theorem in Chapter 6. What property of an electron would an experimenter be resigned to knowing less and less accurately as she determines the electron's precise angular momentum more and more accurately?

3. Carbon-14 has a half-life of 5,730 years.

   a. How long will one atom of C-14 last?

   b. If you start with a sample size of 12 grams of C-14, how many years have gone by when you only have 0.75 grams?

   c. If you start with a sample size of 1 billion atoms of C-14, approximately how many will be left after 5,000 years?

4. Calculate the momentum of an average photon in the visible range.

5. Calculate the wavelength of a 1,400 kg car traveling at 55 mph.

6. Where in the universe can you find an isolated charge of $1.1 \times 10^{-19}$ coulombs?

7. Cs-127 has a half-life of about 30 years. When it undergoes radioactive decay, it does so by beta emission. Write down the nuclear reaction for this decay.

# ASTROPHYSICS

## WHAT YOU WILL LEARN

- Fusion, fission, and the creation of all the elements.
- The life cycle of stars.
- What pulsars, neutron stars, black holes, red giants, and white dwarves are.
- Evidence for the big bang.
- What is meant by the heat death of the universe.
- Dark energy and the expansion of the universe.
- Dark matter and the large-scale structure of the universe.
- How we measure distance in astronomy.

| LESSONS IN CHAPTER 16 | |
| --- | --- |
| • The Composition of Stars | • Dark Energy |
| • Fission and Fusion | • Dark Matter |
| • Stellar Evolution | • The Cosmic Distance Ladder |
| • The Big Bang | |

## 16.1 The Composition of Stars

Until 1925, no one knew what the sun (and, by extension, other stars) was made of. It was commonly assumed that the sun was made of similar stuff to Earth. Thanks to the work of Cecelia Payne that year, we now know that stars (and thus the universe) are primarily made of hydrogen. Subsequently, we have undergone a profound shift in our understanding of the distribution and origin of the elements. Over 99% of all atoms in the universe are hydrogen and helium, as shown in Figure 16.1.

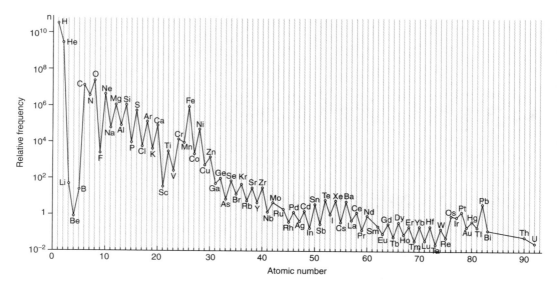

**Figure 16.1** Relative abundance of elements.

As can be seen in the logarithmic graph in Figure 16.1, the abundance of elements varies markedly by element type. Amazingly, this distribution is well understood in terms of stellar evolution. Stars are born (begin to glow via the fusion process), they live (fusing lighter elements into heavier ones), and they die (stop the fusion process by running out of lighter elements).

Without fusion, there is only one element in the universe: hydrogen. Since one proton (and no neutrons) is all that's needed, hydrogen is the primordial element. If the temperature, pressure, and density are high enough, hydrogen will fuse into helium. This fusion process is what is powering our sun and all so-called main sequence stars. As clouds of gaseous hydrogen collect and collapse under their own gravitational attraction, the hydrogen heats up. As it heats up, the mass enters the plasma state: all electrons are free of their host atoms; stars exist in a superheated chaotic dance of charged particles creating intense electric and magnetic forces inside of them. If the mass is insufficient to create enough heat and pressure to trigger fusion, the collection of gas becomes a gas giant planet like Jupiter or Saturn. If the mass is large enough to cause some fusion but not enough to fuse reliably hydrogen into helium, the result is a kind of failed start known as a **brown dwarf**. Stars that are large enough to fuse hydrogen into helium but are smaller than our own star are **red dwarves**. Thought to be the most common type of star, red dwarves glow so faintly and are so small that they are hard to spot. Since they are cooler than most stars, they fuse their hydrogen in a leisurely manner and have the longest life spans of any stars. The life cycle of stars is dictated by their initial mass. The more massive they are, the faster they fuse through their light elements and the sooner they die. Red dwarves are so long-lived that it is thought that none have yet died in the 13.8 billion years that the universe has existed. The most massive of stars fuse through their elements and die in a few million years. Moderate-mass stars like our own sun have life spans of around 10 billion years.

# 16.2 Fission and Fusion

The hydrogen fusion process in main sequence stars is known as proton-proton fusion, which goes through the following sequence of events:

Proton + Proton → Deuterium (H-2, a hydrogen ion containing 1 neutron)

Deuterium + Proton → Helium (He-3, an unstable isotope of helium)

He-3 + He-3 → He-4

At each step in the proton-proton fusion chain, other particles are created to keep charge, momentum, and energy conserved. However in each step, the total mass of the products is *lower* than the total mass of the reactants. This conversion of mass into energy (via $E = mc^2$) is where the energy of the stars come from.

The subsequent fusion of helium into heavier elements is harder to trigger since now there are two protons in each nucleus and therefore even more electrostatic repulsion to overcome for the strong nuclear force to bind the nucleons together. (Indeed, this is the reason such high temperatures and densities are needed in the core of stars even to begin the fusion process at all.) You might think this trend continues straight through the periodic table and that all fusion events are energy releasing. However at some point, there are simply too many protons packed into a single nucleus for the strong nuclear force to bind any more into the nucleus and create a lower-mass state. As shown in Figure 16.2, this happens at element number 26, iron.

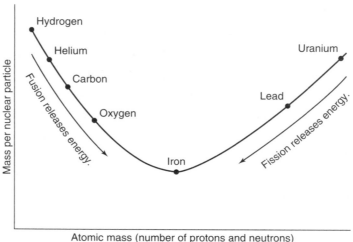

**Figure 16.2** Contrasting fission and fusion releasing energy.

No element higher than iron is created via energy-releasing fusion. These higher-mass nuclei release energy via fission; their products after they break apart have less total mass than the original nuclei. A natural question to ask is how did all the elements heavier than iron come to be created? The answer is that all of these elements are created in supernova events (see "Life Cycle of High-Mass Stars," below). During these brief explosions, enough excess energy is available that the energy-hungry requirements of these heavy-element fusion events can be met.

# 16.3 Stellar Evolution

The life cycle of low-mass stars differs from that of high-mass stars.

## LIFE CYCLE OF LOW-MASS STARS

Main sequence stars heavier than red dwarves eventually fuse most of the hydrogen in their cores and enter a new phase fusion. As the core collapses under its own gravity after the hydrogen fusion is over, it heats up to even higher temperatures. Meanwhile, the outer layers of the star swell up to many times their original size and, in so doing, cool down. This phase of the star's existence is known as a **red giant** for moderate-sized stars like our sun. In the superheated core, helium fusion is now triggered. The first stable fusion process available is the **triple alpha process**: three helium nuclei fuse into a single carbon nucleus. This is where all of the carbon in your body comes from. Some long-dead star underwent the triple alpha process before our sun was even born and eventually donated its contents to the hydrogen cloud that would eventually collapse into our solar system!

For small- to moderate-sized stars like our own sun, the superhot core undergoing the triple alpha process is revealed as a **white dwarf** once the layers of the red giant have faded away. For stars up to a few times our own sun's mass, this is the final state. Our sun will simply not have enough mass at this point to heat up enough to trigger the fusion of carbon into anything heavier. This white dwarf will then spend billions of years of cooling off. Eventually (although the universe is too young for this to have happened to any white dwarf thus far), it is thought a white dwarf will cool enough to no longer emit visible light, at which point it will be called a **black dwarf**.

# LIFE CYCLE OF HIGH-MASS STARS

For stars higher in mass (above 10 solar masses or so), their cores eventually get hot enough to fuse elements higher than carbon. They continue collapsing and creating new layers of hotter and hotter fusion until they have an onion layer style structure with heavier and heavier elements getting fused the deeper you go. In this way, the even-numbered elements up through and including iron are created. This is the reason for both the preponderance of even-numbered elements (since this onion layer–style fusion begins with a mostly helium-based core) and why the elemental abundance for elements beyond iron drops off precipitously. When the star has an inner iron core and gravity attempts to squeeze it down further, there is no standard fusion solution. So there is a different type of collapse. The energy generated blows out the outer layers of the star in an event so bright it can briefly exceed the luminosity of an entire galaxy: a **supernova**. Depending on the mass of the collapsing core, the collapse may result in one of two objects (if anything is left): a neutron star or a black hole.

A **neutron star** is created by combining all of the electrons and protons and by eliminating the charges that keep the nucleons separated. The resulting stellar-weight nucleus is the densest substance in the universe. A typical neutron star has more mass than our current sun packed into a diameter of a few dozen kilometers. As stars collapse, they must conserve their angular momentum, and thus neutron stars are spinning quite rapidly. If their remaining magnetic poles stir up periodic bursts of electromagnetic radiation, they are known as **pulsars**. Neutron stars are no longer generating heat and are held up by the Pauli exclusion principle. The same quantum rule that prevents two electrons from sharing space also applies to neutrons. (Actually, the exclusion principle applies to all fermions. This is why matter, which is made up of fermions, takes up space, whereas their interaction forces—made of bosons—do not).

A **black hole** is the ultimate victory of gravity over matter and is the end game for the largest of stars. In their case, the collapse of the iron core does not stop at the neutron level but, rather, collapses all the matter to the point that a possible solution once thought to be absurd to Einstein's general relativity is realized: a curvature of space-time so severe that not even light can escape and inside of which time dilation is so pronounced that time ceases to exist. When first shown to be a solution to his equations, even Einstein didn't believe that such objects would exist in reality. In fact, the term *black hole* was conjured up in mockery of the concept. However, we have now observed several stellar mass black holes (thought to have been created in the process just described) as well as some supermassive black holes whose origins are still being debated. The first supermassive black hole observed was at the center of our own galaxy. Since black holes do not emit any radiation, they are "seen" by the motion of stars around them or through their gravitational lensing effects on light passing nearby. The sphere inside of which nothing can escape is characterized by its **Schwarzschild radius** (or, more colloquially, the event horizon of the black hole). Objects outside of the event horizon are free to revolve or otherwise gravitationally interact with the black hole normally.

## 16.4 The Big Bang

One of the natural solutions to Einstein's general theory of relativity shows a natural expansion of space-time. Prior to observational evidence compiled by Edwin Hubble in the late 1920s, however, this was considered a defect of the theory and was not taken seriously. In fact, prior to Hubble's observations, it was not known whether the Milky Way galaxy was the only galaxy or if there were others. Hubble's careful measurements of distance and the red shift of the light from those most distant stars demonstrated two things at once: (1) other galaxies exist and (2) most of them are moving away from us. Not only are most galaxies moving away from us (all, except for some close enough to be gravitationally attracted to us, are, in fact, moving away from us), but they are moving away from us faster the farther away they are. This surprising observation makes sense only if we live in an expanding universe. Since every cubic centimeter is expanding, the more centimeters there are between two objects, the faster these objects will be carried apart by the expansion of space. Objects close enough together to be bound electrically or gravitationally easily overpower this expansion. However, for objects very far apart, the expansion can be mind-blowingly fast. There are galaxies so far away that there is so much space expanding between us; they are being carried away from us faster than the speed of light so we will never see them! The last objects we can currently see are at the edge of what we call the **observable universe**.

The idea of the big bang comes from winding clocks backward. Since we are expanding now, if you play time backward, the universe will contract. At some point in the distant past, the entire universe was collected in a tiny volume. The idea of this beginning of space and time expanding into the currently sized universe is the basis of the term *big bang*. In fact, we are, have been, and will continue to expand so the big bang is happening all the time, all around us. Since we are talking about the expansion of space itself, there is no location to point to as the origin of the expansion. (We can point backward in time, but that is the best we can do.) Subsequent to Hubble's measurements of the expansion, there have been several other experimental confirmations of the expansion and finite age of the universe:

- The existence of a predicted **cosmic microwave background** from a time in the distant past when the universe first cooled off enough to allow light to escape the original plasma soup

- The different state of a much younger universe as we look farther and farther away from Earth (the farther away you look, the longer it has taken light to reach us and thus you are looking back in time)

- The lack of any objects that would require more than 14 billion years in order to exist (such as white dwarves that have faded into black dwarves or red dwarves that have fused all their hydrogen)

The idea that time has an actual beginning some 13.8 billion years ago is philosophically interesting in and of itself. However, consider this: if given enough time, the universe will be spread so thinly that all the hydrogen that could possibly have been gravitationally bound up into stars will have already been fused. No more stars will be born. Just as there must have been a dark era near the beginning of the universe before the first stars were born, there will probably be a dark fate in the future when no more stars will shine and everything will wind down to a cold halt in an ever-expanding universe. This idea is called the **heat death** of the universe.

# 16.5 Dark Energy

Prior to the 1990s, many astrophysicists thought that the expansion of the universe should be slowing down. Their reasoning was that all the gravitational attraction should be working against the expansion rate. Careful measurements of the expansion of the universe beginning at that time and continuing to this day have found the opposite to be true: the universe is accelerating in its expansion! Since we currently have no mechanism for this acceleration of expansion and there is no known means of creating the energy needed for this expansion, dark energy is the label given to the source of this accelerating expansion.

# 16.6 Dark Matter

Since the early days of the discovery of galaxies, astrophysicists noticed that there seems to be more gravity holding things together than could be accounted for by the mass known to be present. Two of the most famous pieces of evidence for this are the rates of revolutions in stars about their galactic centers and gravitational lensing. In the 1970s, Vera Rubin observed that stars in the outer portions of galaxies were revolving more quickly than predicted (see Figure 16.3).

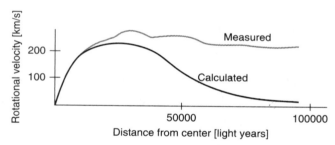

**Figure 16.3** Predicted vs. measured rotational velocities of stars.

Either our understanding of gravity is wrong or else there is more matter that we cannot see spread evenly throughout these galaxies. Further evidence for the existence of the so-called dark matter came when looking at the gravitational lensing around galaxies. The amount of gravitational lensing as predicted by general relativity is proportional to the mass present. There is simply more lensing than would be predicted by all the normal matter thought to be present. Years of careful searching has not revealed what dark matter is, but several conventional ideas can be ruled out. The missing matter is not made of particles in our standard model or their antimatter counterparts, and the missing matter is not made of black holes. Furthermore, the large-scale structure of the universe can be explained only by the presence of dark matter, a substance that interacts only gravitationally. The large-scale structure of the universe refers to the fact that galaxies are grouped into clusters and superclusters with giant voids embedded in the otherwise evenly spread matter density. These clusters and superclusters are themselves forming a weblike structure in the universe. As the universe expands globally, locally the galaxies come together to draw the tendrils of the cosmic web tighter. In computer simulations of the evolution of this large-scale structure, massive amounts of dark matter are required in order to produce the observed results.

# 16.7 The Cosmic Distance Ladder

How do we know how far away the stars are? This seemingly innocent question turns out to be a very difficult one to answer. In fact, there is no single answer. Rather, scientists have determined the distances to various astronomical objects by using a hodgepodge of techniques developed over the past hundred years. Each time we look farther away, a new technique is needed. Oftentimes, the new method is dependent on the previous one. For this reason, the collection of techniques is known as rungs on the cosmic distance ladder.

# RUNG 1: RADAR RANGING

Radar is the light version of echolocation: bounce a signal off something and measure the time. With radio waves, the speed is the speed of light. If the object is big enough and close enough, we can detect the echo with a radio telescope and determine the distance by the time delay. This technique works for the shorter distances within the solar system and is the primary means by which we have accurate measurements of distances between the planets and to the sun. However, once you start trying to find the distances to objects in the outer solar system or even to small objects within the solar system, there simply is not enough signal reflected back to Earth to be detectable.

# RUNG 2: PARALLAX

To measure the distances to the closest stars, astronomers used the same technique that most animals use to give us depth perception: parallax. The two eyes send the brain two slightly different images of the same object. The difference in viewing angle from one eye to the other allows the brain to do a rough calculation of distance. A geometry teacher would say we are using the trigonometry of triangles to calculate distance where one side of the triangle is the known distance between our eyes. In astronomy, measurements of the same star are made when Earth is on either side of the sun. These two locations of Earth are like our two eyes with the separation (or base of the triangle) being the orbital diameter of the Earth about the sun. In this way, we can directly calculate the distance to start within 1,000 light-years. Beyond 1,000 light-years or so, the parallax triangle has too narrow a base to obtain measurable differences in angle from the two sides of our orbit.

## RUNG 3: SPECTROSCOPIC PARALLAX

Enough stars are within 1,000 light-years of Earth (our local galactic neighborhood but far shy of the 100,000 light-year diameter of our galaxy) that we have built up a catalog of star types and brightnesses. The type of star is determined by doing spectroscopy on the light from the star, which then tells us lots of information about the star: composition, temperature, etc. Since we know the apparent brightness of the star (how bright it appears to us here on Earth) and the distance to these stars from the parallax method, we can calculate these stars' absolute brightness (how bright they would appear if we were right next to them).

For the next rung (up to about $10^6$ light-years), we can use this information from our catalog of known stars as follows. Run the spectroscopy of a star that is too distant to measure the distance to via parallax. Next, match this new star to a known star of the same type, and assign it the same absolute magnitude of brightness. Finally, by taking the difference between the absolute and apparent brightness, we can determine the distance. As light travels out from a star in all directions, the apparent magnitude falls off as a well-defined function of distance. However, once the star is too far away to do detailed spectroscopy, this technique no longer works.

## RUNG 4: CEPHEID VARIABLES

Throughout the early 1900s, a special type of star became increasingly important in determining distance. Cepheid variables are variable in their brightness. After Henrietta Swan Leavitt discovered that their period of variation was related to their absolute brightness, these stars became extremely useful since their periodicity was easy to identify, even at distances beyond which spectroscopy of individual stars can be done. By taking the difference between the apparent magnitude of distant Cepheid variables as seen from Earth and their absolute magnitudes as determined by their periodicity, their distance from Earth could be calculated. Objects that we know the absolute magnitude of and thus can be used to determine distance in this way are called *standard candles*.

In 1924, Hubble used Cepheid variables observed in the Andromeda nebula to determine that the distance to this so-called nebula clearly placed it outside of our own Milky Way galaxy (2.5 million light-years away!). This was the first proof that our galaxy is one of many, and the Andromeda was properly classified as its own galaxy. In 1929, Hubble went on to use the newly calculated distances to several galaxies along with their measured velocities (as determined by the red shifting of their light) to offer the first proof that we are living in an expanding universe.

## RUNG 5: SUPERNOVA 1A

As useful as they are, in the most distant of galaxies (greater than $10^8$ light-years away), even the variation of a Cepheid variable cannot be distinguished from the light of all the stars in those galaxies. What is bright enough? A supernova can be seen even in the most distant of galaxies. Many supernovas are from high-mass stars collapsing. Their brightness is directly related to their mass, thus not making for good standard candles. There is one type, however, that is much more reliable: supernova 1a. This special type of supernova makes for a good standard candle since it happens when a low-mass star (a white dwarf) crosses the mass threshold to go supernova from below. If a white dwarf (which is normally not heavy enough to go supernova by definition) is in a position to accumulate more mass (from a companion star), it can eventually increase its mass to the critical 1.4 times the sun's mass that will trigger a supernova. Since all these supernova 1a are triggered by the same mass, they all have the same brightness: the brightest of all standard candles. It is precisely these standard candles that were used to map the distances to the most distant of galaxies that led to the discovery of the accelerating expansion of the universe. This acceleration of the expansion is fueled by means unknown that is now called dark energy.

# VECTORS

## Concepts

### WHAT ARE VECTORS?

Mathematics is the language of physics. When physicists seek to describe the universe we live in, they most frequently describe the properties they see mathematically. Two basic types of mathematical entities are commonly used in physics: **scalars** and **vectors**.

**Scalars** are simply numbers. For example, 55 mph or 12 inches are scalar quantities because they are numbers without direction. In the first example, the number 55 tells you the speed of something in miles per hour. In the second example, the number 12 tells you the length of something in inches. Neither example, however, tells you in what direction the speed or the length is pointed.

Note that physicists always insist on attaching units to their numbers. A length of 12 is meaningless unless you also know the units: 12 *what*? Inches? Feet? Miles? Meters? Kilometers? Light-years?

**Vectors** are numerical entities that also specify direction. Examples of vector quantities include 55 mph due north or 12 inches upward from the ground. Like scalars, vectors must include units that give the number a physical meaning. Unlike scalars, vectors also include a direction.

A vector's direction allows the vector to be represented graphically. A vector may be pictured as an arrow. The length of the arrow (the scalar quantity) is called the *magnitude* of the vector. The direction, or angle, in which the arrow is pointed is the direction of the vector. Note that mathematically, two vectors with the same magnitude and direction are equivalent even if they are located in different places. For example, 55 mph due north is the same vector in Chicago as it is Boston.

Why bother with this distinction between scalars and vectors? Frequently, scalars are simply not good enough; scalars do not provide enough information. Imagine having driving directions such as the following: "Go 55 mph for one hour and then go 55 mph for another hour." Where will such directions take you? There is simply not enough information provided to answer that question. You might end up 110 miles away from your original location if you travel in the same direction both hours. Instead, you might end up exactly where you started if you travel in exactly the opposite direction during the second hour. Of course, you might also end up somewhere else if the directions are different but not exactly opposite. To know the answer for sure, you would need to be told the direction in which to travel for each hour.

## GRAPHICALLY ADDING VECTORS

So how does graphically representing vectors as arrows help us? To add any two vectors, we can simply draw the two arrows to scale, placing the tail of the second arrow at the point of the first, as shown in Figure A.1. The vector sum is the new arrow that can be drawn from the tail of the first arrow directly to the point of the second arrow.

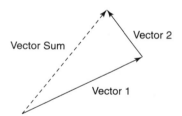

**Figure A.1** Adding vectors graphically.

In Figure A.1, the dashed arrow represents the vector sum of the two solid arrows. The length of the arrow representing the vector sum is the magnitude of the sum of the two vectors. The sum of two or more vectors is also called the *resultant*.

If two vectors can be added graphically, we might expect to be able to subtract vectors graphically. Consider subtraction as simply addition using a negative sign. For example: $5 - 3 = 5 + (-3)$. A minus sign in front of a vector simply reverses its direction; the arrow is rotated by exactly 180°. Building upon the vector addition example above, we can subtract Vector 2 from Vector 1 as shown in Figure A.2.

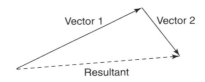

**Figure A.2** Subtracting vectors graphically

# Math

## NOTATION

Unfortunately, there are many equivalent notations used to indicate vectors.

- One method frequently adopted by textbooks is to print vector quantities in bold. Thus, $A$ represents a scalar quantity, whereas **A** represents a vector quantity.

- More commonly, a vector quantity is identified by placing a tiny arrow above the variable. So $\vec{A}$ represents a vector.

- A *unit vector* (a vector with a magnitude, or length, of exactly 1 unit) is identified by a carat, or hat, in the notation, for example, $\hat{A}$.

- The magnitude of a vector is always positive and is commonly abbreviated as $|\vec{A}|$.

## VECTOR COMPONENTS

The graphical method of adding and subtracting vectors, though useful for visualizing the vectors and their interaction, is cumbersome in practice. It is much easier to break down a vector into its *components*. Vectors can be broken down into two components. One describes how far "over" (typically in the $x$-direction) and the other describes how far "up" (typically in the $y$-direction) we must travel to get from the tail end of the vector arrow to the tip of the arrow.

Mathematicians frequently express vector components in *coordinate notation*. For example, (3, 4) represents a vector arrow whose tip can be found at 3 units in the x-direction and 4 units in the y-direction from the origin (0, 0). By using our knowledge of trigonometry, we also know that the vector has a length, or magnitude, of 5 and is pointed approximately 53° above the horizontal x-axis.

$a^2 + b^2 = c^2$

If we call this vector $\vec{A}$, *component notation* is an alternate method for expressing the vector:

$$A_x = 3$$

$$A_y = 4$$

Such notation can be read as "*A* sub *x* is equal to 3 and the y-component of vector *A* is 4."

True *vector notation* makes use of unit vectors. When using unit vectors, the same vector $\vec{A}$ can be expressed as follows:

$$\vec{A} = 3\hat{x} + 4\hat{y}$$

This notation can be read as "the vector *A* is 3 units in the x-direction and 4 units in the y-direction."

Note that if the vector lies anywhere other than in the first quadrant (where both x- and y-components are positive), one or both components must be negative. Remember that a negative sign is simply a reversal of direction. Table A.1 shows different vectors, all with the same magnitude but pointed in different directions.

**TABLE A.1  VECTORS WITH THE SAME MAGNITUDE BUT DIFFERENT DIRECTIONS**

| Coordinate Notation | Component Notation | Vector Notation | Vector Quadrant |
|---|---|---|---|
| (3, 4) | $A_x = 3$ <br> $A_y = 4$ | $\vec{A} = 3\hat{x} + 4\hat{y}$ | First quadrant |
| (−3, 4) | $A_x = -3$ <br> $A_y = 4$ | $\vec{A} = -3\hat{x} + 4\hat{y}$ | Second quadrant |
| (−3, −4) | $A_x = -3$ <br> $A_y = -4$ | $\vec{A} = -3\hat{x} + (-4\hat{y})$ | Third quadrant |
| (3, −4) | $A_x = 3$ <br> $A_y = -4$ | $\vec{A} = 3\hat{x} + (-4\hat{y})$ | Fourth quadrant |

How do we convert component notation to polar notation, and vice versa? We use trigonometry. The $x$- and $y$-components ($V_x$ and $V_y$) of a vector are the legs of a right triangle with the vector arrow itself being the hypotenuse (i.e., $r = |\vec{V}|$). We can derive polar components using all of the basic trigonometric relations we've learned in mathematics:

$$V_x = |\vec{V}|\cos\theta$$

$$V_y = |\vec{V}|\sin\theta$$

$$\theta = \tan^{-1}\left(\frac{V_y}{V_x}\right)$$

We must be careful with our signs for the components and which angle we are using with these trig relationships. Figures A.3 through A.6 show vectors in each quadrant.

$$V_x = +|\vec{V}|\cos\theta$$

$$V_y = +|\vec{V}|\sin\theta$$

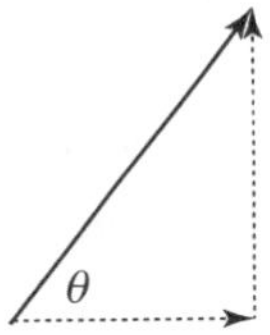

**Figure A.3** Qudarant I vectors.

$$V_x = -|\vec{V}|\cos\theta$$

$$V_y = +|\vec{V}|\sin\theta$$

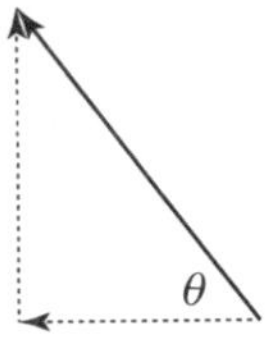

**Figure A.4** Quadrant II vectors.

$$V_x = -|\vec{V}|\cos\theta$$

$$V_y = -|\vec{V}|\sin\theta$$

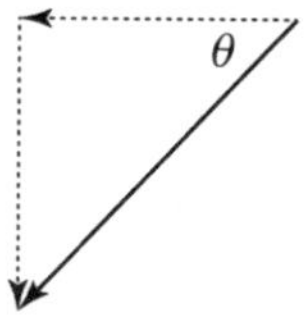

**Figure A.5** Quadrant III vectors.

$$V_x = +|\vec{V}|\cos\theta$$

$$V_y = -|\vec{V}|\sin\theta$$

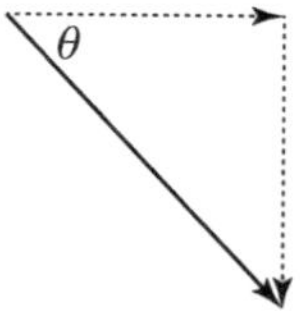

**Figure A.6** Quadrant IV vectors.

Two vectors are equal if and only if they have the same magnitude and the same direction. Graphically, this means that the vectors' arrows must be the same length and pointed in the same direction. In regard to vector components, this means that two vectors are equal only when each of their components are equal.

If given a vector equality, e.g., $\vec{A} = \vec{B}$, there are as many scalar equalities as there are components in the vectors:

$$A_x = B_x$$

$$A_y = B_y$$

etc.

# MATHEMATICALLY ADDING VECTORS

When adding vectors, rather than drawing a series of arrows on graph paper and graphically determining the resultant, we can simply add the components of the vectors.

For example, given vectors $\vec{A}$, $\vec{B}$, and $\vec{C}$, suppose we are interested in $\vec{D} = \vec{A} + \vec{B} + \vec{C}$. The resultant $\vec{D}$ is defined by the following:

$$D_x = A_x + B_x + C_x$$

$$D_y = A_y + B_y + C_y$$

$$|\vec{D}| = \sqrt{D_x^2 + D_y^2}$$

How many components does it take to describe a vector completely? As a general rule, a vector requires as many components as there are dimensions for it to occupy. Theoretically, this means a vector can be made up of any number of components. For practical purposes, however, most physics problems are two-dimensional $(x, y)$ with some three-dimensional $(x, y, z)$ problems when required.

# AN EQUIVALENT DESCRIPTION

Let's consider vectors in the familiar two-dimensional plane. According to the component method, a two-dimensional vector requires two components, $x$ and $y$, to describe it fully.

We can, however, also express the vector in such a way as to describe the graphical arrow notation. Again, there are two components. In this case, one is the length of the arrow (the magnitude of the vector), and the other is the angle the vector makes with respect to some reference direction. Unless another reference is specified, the standard angle is given as a counterclockwise rotation from the positive $x$-axis. If both the magnitude and the angle are given, the vector is considered complete. Mathematicians sometimes refer to this magnitude-angle format of vectors as *polar notation*. It is expressed as $(r, \theta)$, where $r$ is the magnitude of the vector and $\theta$ (read "theta") is the angle. For example, the vector written in component notation as (3, 4) can also be written in polar notation as (5, 53°). If it is a vector in three dimensions, then three components are needed: magnitude and two angles $(r, \theta, \Phi)$, where $\Phi$ (read "phi") is angle of elevation above the $xy$-plane.

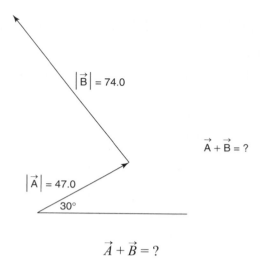

$$\vec{A} + \vec{B} = ?$$

$$\vec{A} + \vec{B} = ?$$

## SOLUTION

Step 1: Choose the orientation of axes (basic vectors).

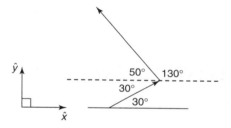

Step 2: Break each vector into its components.

$$\vec{A} = 47\cos30°\hat{x} + 47\sin30°\hat{y}$$

$$\vec{B} = 74\cos130°\hat{x} + 74\sin130°\hat{y}$$

Step 3: Perform the required operations component by component.

$$\vec{C} = \vec{A} + \vec{B} = C_x\hat{x} + C_y\hat{y}$$

$$C_x = A_x + B_x = 47\cos30° + 74\cos130° = 40.7 - 47.6 = -6.87$$

$$C_y = A_y + B_y = 47\cos30° + 74\cos130° = 23.5 - 56.7 = 80.2$$

$$\vec{C} = -6.87\hat{x} + 80.2\hat{y} \qquad |\vec{C}|^2 = C_x^2 + C_y^2 = (-6.87)^2 + (80.2)^2 = 6.47 \times 10^3$$

$$|\vec{C}| = 80.5 \quad \tan\theta = \frac{C_y}{C_x}$$

$$\tan\theta = \frac{80.2}{-6.87} = -11.7$$

$$\theta = \arctan(-11.7) = \tan^{-1}(-11.7) = -85.1°$$

Although the calculator provides a solution of −85.1° for θ, *this answer is incorrect!* At this point in your problem solving, you must stop and *think* about whether to accept the calculator's answer.

The calculator gives the smallest angle that is a *possible* solution. In general, all tangent functions have two possible answers. The first and third quadrants share a solution set, as do the second and fourth quadrants. To derive a second solution based on the calculator's solution, recall that the tangent function repeats itself every 180°. So simply add 180° to your calculator's answer when necessary.

In this example, we know that the resultant vector must lie in the second quadrant (negative *x*, positive *y*). The calculator's answer falls in the fourth quadrant (positive *x*, negative *y*). To derive another solution for $\tan^{-1}(-11.7)$, we add 180° to the calculator's answer of −85.1°.

$$180 + (-85.1) = 94.9$$

$$\theta = 94.9°$$

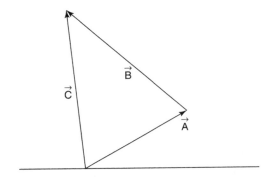

Step 4: Check the answer for *units*, *significant figures*, *proper scientific notation*, and *physical plausibility* if solving a real-life problem.

Since the problem provided three significant digits, make sure your answer has three significant digits as well.

Check that the solution makes sense.

# MULTIPLYING VECTORS

There are two ways to multiply vectors. One method is called a **dot product**. The result is a scalar, which indicates to what extent the two vectors are working along the same direction. A positive sign indicates the vectors are working together; a negative sign means they are mostly in opposition. Their dot product value can vary from zero (the vectors are at right angles; *orthogonal*) to a maximum of the algebraic product of their magnitudes (*collinear*).

Just as with the components of vectors, there are two ways to find the products of vectors in each multiplication method, one by components and the other by magnitude and angle. In this case, however, the key angle is not the rotation from the *x*-axis but, rather, the included angle between the two vectors when placed tail to tail.

Component method:

$$\vec{A} \cdot \vec{B} = A_x B_x + A_y B_y + A_z B_z$$

Included angle (θ) method:

$$\vec{A} \cdot \vec{B} = |\vec{A}||\vec{B}|\cos\theta$$

The other method to multiply vectors is called the **cross product**. Its result is a vector whose direction indicates the axis of rotation that the two vectors form and whose magnitude indicates how well the two would cause each other to rotate. The cross product is used with torques, angular momentum, and magnetism.

Component method:

$$\vec{A} \times \vec{B} = (A_x B_y - A_y B_x)\hat{z} + (A_x B_y - A_y B_x)\hat{y} + (A_x B_y - A_y B_x)\hat{x}$$

Included angle (θ) method:

$$|\vec{A} \times \vec{B}| = |\vec{A}||\vec{B}|\sin\theta$$

In this method, the direction of the resultant is given by the right-hand rule. Put your right hand's thumb in the direction of $A$ and your first finger in the direction of $B$. Now extend your middle finger so that it is at right angles to both fingers. The direction of your middle finger is the direction of the resultant.

# Physics

Why all the fuss about vectors? Some physical properties of objects are innately scalars, and some are innately vectors. For example, many physical laws can be modeled as an interaction of vectors. Because scalars and vectors are treated differently, you must be careful to determine whether you are dealing with scalar or vector quantities.

Table A.2 lists commonly found scalar and vector quantities. Related quantities are listed in the same row.

### TABLE A.2 COMMON SCALAR AND VECTOR QUANTITIES

| Scalars | Vectors |
| --- | --- |
| Distance | Displacement |
| Time | |
| Speed | Velocity |
| | Acceleration |
| Mass | |
| | Force |
| Energy | |
| | Momentum (linear and angular) |
| Charge | |
| Electric potential | Electric field |
| | Magnetic field |
| Temperature | |

Note that usually there will be units (such as mph or inches) associated with each property value. When dealing with vectors, the magnitude of the vector and of each component of the vector all must have the same unit!

Note also that in physics, when asked to solve for a vector variable, you *must* provide complete information—either in component notation with $x$ and $y$ coordinates or in polar notation with a magnitude and direction. Otherwise, your answer is considered ambiguous.

# Appendix A Review Exercises

**Given two vectors of magnitudes 7 units and 5 units, what are the minimum and maximum magnitudes of all the possible ways they could be added together?**

1. If you have a vector of magnitude 5 units straight up, another one of 6 units to the right, an additional one of 7 units to the left, and another one of 5 units down, what is the resultant vector of these four vectors?

2. Given a vector of magnitude 10 units at an angle of 145 degrees, what are the horizontal and vertical components of this vector?

3. Given a vector with an $x$-component of +3 units and a $y$-component of –4 units, what is the magnitude and direction of the vector?

4. You are given the vectors (2, 5) and (–7, 8).

   a. Graphically (using graph paper, a ruler, and a protractor) add these vectors.

   b. Graphically (using graph paper, a ruler, and a protractor) subtract the vector (–7, 8) from the vector (2, 5).

   c. Check your results for parts (a) and (b) using algebra and trigonometry.

5. Given:

$$A_x = 9, A_y = 2 \qquad B_x = -4, B_y = -4 \qquad C_x = 5, C_y = -3$$

   a. What is $\vec{A} + \vec{B} + \vec{C}$? Give your answer in *both* component $(x, y)$ and polar (magnitude-direction) notation.

   b. What is $\vec{A} + \vec{B} - 2\vec{C}$? Give your answer in *both* component $(x, y)$ and polar (magnitude-direction) notation.

# UNITS

## SI Units

Formerly known as the metric system, the entire International System of Units (SI units) are based on only seven base units, as shown in Table B.1. All units used can be reformulated as some combination or transformation of these units, as shown in Table B.2. Originally, many of the base units were defined in terms of exemplars created for the SI system itself. Through the hard work of many scientists to measure and agree with high precision about the fundamental constants in science, as of 2019, all seven of the base units have now been redefined. These definitions are in terms of fundamental constants and natural phenomena (such as the speed of light, the charge on an electron, the Boltzmann constant, etc.) that are, in turn, held to be precise.

### TABLE B.1  BASE SI UNITS

| Quantity | Unit Name | Abbreviation | Defined By |
|---|---|---|---|
| Length | Meter | m | Speed of light |
| Mass | Kilogram | kg | Planck's constant |
| Time | Second | s | The frequency of a certain atomic transition in Cs-133 |
| Electric current | Ampere | A | Elementary unit of charge |
| Temperature | Kelvin | K | Boltzmann constant |
| Amount of substance | Mole | mol | $6.02214076 \times 10^{23}$ |
| Luminous intensity | Candela | cd | Specific power per square radian (steradian) of a specific frequency of light |

## TABLE B.2  DERIVED SI UNITS COMMONLY USED IN PHYSICS

| Quantity | Unit Name | Abbreviation | Expression in Other SI Units |
|---|---|---|---|
| Linear velocity | | | m/s |
| Linear acceleration | | | m/s$^2$ |
| Force | Newton | N | kg • m/s$^2$ |
| Momentum | | | kg • m/s |
| Impulse | | | N • s = kg • m/s |
| Angular velocity | | | rad/s |
| Angular acceleration | | | rad/s$^2$ |
| Torque | | | N • m |
| Angular momentum | | | kg • m$^2$/s |
| Moment of inertia | | | kg • m$^2$ |
| Spring constant | | N/m | kg/s$^2$ |
| Frequency | Hertz | Hz | cycles/s |
| Pressure | Pascal | Pa | N/m$^2$ = kg/(m • s$^2$) |
| Work, energy | Joule | J | N • m = kg • m$^2$/s$^2$ |
| Power | Watt | W | J/s = kg • m$^2$/s$^3$ |
| Electric charge | Coulomb | C | A • s |
| Electric potential | Volt | V | J/C = kg • m$^2$/(A • s$^3$) |
| Resistance | Ohm | Ω | V/A = kg • m$^2$/(A$^2$ • s$^3$) |

Note: Rad (radians) and cycles are not true units. However, angles and frequencies are sometimes measured in other units and thus specifying them, although formally not required, is useful.

# Dimensional Analysis

When transforming or manipulating units, formal algebraic rules apply. This is known as dimensional analysis. The key to dimensional analysis is that units follow the same rules as numbers for multiplying and dividing:

$$\frac{(2 \text{ cm})(3 \text{ cm})(4 \text{ cm})}{6 \text{ s}} = 4 \text{ cm}^3/\text{s}$$

For unit conversions, the key is to change a conversion to a fraction equal to 1:

$$2.54 \text{ cm} = 1 \text{ inch}$$

becomes either

$$\frac{2.54 \text{ cm}}{1 \text{ inch}} = 1 \quad \text{or} \quad \frac{1 \text{ inch}}{2.54 \text{ cm}} = 1$$

Why does dimensional analysis work? You can multiply anything by 1 as often as you like without changing the actual value of the number (or unit).

You no longer have to guess whether to multiply or divide by the conversion factor when changing units. Simply take the starting value and multiply by our fancy 1, taking care to choose the fraction that is stacked in such a way to cancel the unit you are trying to get rid of.

EXAMPLE   Change the following measurement into inches:

$$17 \text{ cm}$$

**SOLUTION**

You must choose between using (2.54 cm/1 inch) or (1 inch/2.54 cm) to convert. Choose the second fraction since you will have centimeters in the denominator, allowing the centimeters to be canceled in the original measurement:

$$17 \text{ cm} \times \left( \frac{1 \text{ inch}}{2.54 \text{ cm}} \right) = \left( \frac{17 \text{ cm}}{1} \right) \times \left( \frac{1 \text{ inch}}{2.54 \text{ cm}} \right) = 6.69 \text{ inches}$$

Notice this has not changed the value of 17 cm but only the units since all you did was multiply by 1.

**EXAMPLE**  Change 55 miles/hour into km/s.

> **SOLUTION**
>
> Since 1.6 km = 1 mile and 3,600 s = 1 hour, you have several conversion fractions to choose from:
>
> $$\frac{1.6 \text{ km}}{1 \text{ mile}} \qquad \frac{1 \text{ mile}}{1.6 \text{ km}} \qquad \frac{1 \text{ hour}}{3,600 \text{ s}} \qquad \frac{3,600 \text{ s}}{1 \text{ hour}}$$
>
> You should choose the first and the third fractions for this conversion because those are the ones that will cancel out mile and hour:
>
> $$\left(\frac{55 \text{ miles}}{1 \text{ hour}}\right) \times \left(\frac{1.6 \text{ km}}{1 \text{ mile}}\right) \times \left(\frac{1 \text{ hour}}{3,600 \text{ s}}\right) = 0.024 \ \frac{\text{km}}{\text{s}}$$

# Universal Units?

Scientists have developed base units such as length (meters), time (seconds), and mass (kg) based on our own historical needs. Indeed, the metric system is sometimes abbreviated as the MKS system for the preeminence of these three units. Our needs for measurement are clearly informed by our own unique scales of interaction with the universe. However, aliens on other planets would presumably have their own, different set of basic units. However, the laws of physics should be the same! How could we go about a unit transformation between our units and theirs without physically comparing them? First, let's take a look at three universal constants we have "discovered."

1. When we measure the "strength" of gravity in our SI units, the universe gives us the following gravitational constant:

$$G = 6.67 \times 10^{-11} \text{ N} \cdot \text{m}^2/\text{kg}^2$$

Note that the extremely small value of this gravitational constant implies that this force is an extremely weak one on our scale.

2. When we measure electricity and magnetic forces in our own units, the universe gives us the speed of light:

$$c = 2.99 \times 10^8 \text{ m/s}$$

Note that the extremely high value for the speed of light in our units is the reason we do not notice relativistic effects in everyday light.

3. The quantum world gives us another fundamental constant, called Planck's constant:

$$h = 6.63 \times 10^{-34} \text{ J} \cdot \text{s}$$

Note that the extremely small value of this unit indicates why we do not notice quantum effects in our everyday life.

Since these measurements depend on the laws of the universe rather than on our own scales of interaction, it is an interesting exercise to use a combination of these universal constants ($G$, $c$, and $h$) to come up with universal units of length, time, and mass. Named after Max Planck who originally proposed them, these units are known as Planck length, Planck time, and Planck mass:

$$\text{Planck length} = \sqrt{\frac{hG}{c^3}}$$

$$\text{Planck mass} = \sqrt{\frac{hc}{G}}$$

$$\text{Planck time} = \sqrt{\frac{hG}{c^5}}$$

In this manner, all of the units needed to describe the known laws of physics can be rewritten in Planck units.

# FORMULAS

## General Physics Problem-Solving Procedure

The following shows a general problem-solving procedure for physics problems and gives reasons *why* we are using these techniques.

1.  Draw a sketch of the situation after reading the problem.

    Reason: A picture is worth a 1,000 words—you must activate the visualization part of your brain.

2.  Make an informed guess at a reasonable answer.

    Reason: Guessing forces you to think about the problem conceptually before getting bogged down in the math, gives you a baseline for a reasonable answer, and forces your brain to switch to thinking mode.

3.  Write down the given information in terms of physics variables. Make sure you adopt a standard sign convention, and stick to your +/– choice for the rest of the problem. Write it directly on your sketch. There may be additional information that is not given in the problem explicitly but that you should know (e.g., acceleration due to gravity, vertical velocity at the highest point, mass of a proton, and velocity when at rest).

    Reason: This discipline forces you to recognize all the given information, even the implied information, and forces you to be explicit with sign conventions.

4.  Pick a general physics equation that contains the variable representing the information requested. After writing down the general form, solve for the unknown algebraically.

    Reason: By going straight to the general equation, you will make no careless assumptions. By solving algebraically without using numbers, you separate your math steps from your calculator/calculation skills and make your mistakes easier to isolate.

5.  Plug in the given information. Remember, you need one equation for each unknown to get a solution. So don't be shy about going back into the physics toolbox and grabbing another equation. Frequently, there is an additional problem-specific equation hidden in the language of the problem itself (e.g., "the total time for the trip was 2 hours").

    Reason: Often you will need to solve for unrequested information by using a different formula to solve for the answer to the problem!

6.  Check your answer:

    *   Against your guess for plausibility—if they are not in the same ballpark, something is wrong, either with your initial assumptions or your math;

    *   To make sure the units are correct and explicit;

    *   To make sure the direction is correct and explicit (if the answer is a vector);

    *   To make sure you have the correct number of significant figures for the problem and have used scientific notation for large or small numbers.

    Reason: Even if you are out of time, acknowledging a ridiculous answer will endear you to the grader!

# Commonly Used Physics Equations

Tables C.1 through C.9 list common physics equations. Each table presents different types of equations.

## TABLE C.1  KINEMATICS EQUATIONS

| | |
|---|---|
| $\Delta \vec{d} = \vec{v}_i + \dfrac{1}{2}\vec{a}t^2$ | Displacement under constant acceleration |
| $\vec{v}_f = \vec{v}_i + \vec{a}t$ | Changing velocity under constant acceleration |
| $\Delta \theta = \omega_0 t + \dfrac{1}{2}\alpha t^2$ | Angular rotation under constant angular acceleration |
| $a_c = \dfrac{v^2}{r}$ | Centripetal acceleration; acceleration needed to take a turn of radius $r$ at velocity $v$ |
| Average speed = distance/time | Scalars; no direction; no negatives |
| $\vec{v}_{avg} = \dfrac{\Delta \vec{d}}{\Delta t}$ | Vectors/slope on direction vs. time graphs |

## TABLE C.2  DYNAMICAL EQUATIONS

| | |
|---|---|
| $\vec{a} = \dfrac{\vec{F}_{net}}{m}$ | Acceleration is caused by total force |
| $\tau = \lvert \vec{r} \rvert F_\perp = r_\perp \lvert \vec{F} \rvert = \lvert \vec{r} \rvert \lvert \vec{F} \rvert \sin\theta$ | Torque is force applied over a lever arm |
| $F_g = \dfrac{GMm}{r^2}$ | Gravitational force (at Earth's surface) |
| $I = \Sigma mr^2$ | Rotation inertia; moment of inertia |
| $\vec{\tau}_{net} = I\vec{\alpha}$ | Total torque causes angular rotations |
| $F_f \le \mu N$ | Sliding friction is proportional to the normal force between the two surfaces |
| $\vec{F} = -k\vec{x}$ | Springs or other restorative forces exert forces proportional to their displacement from equilibrium |
| $\vec{F}\Delta t = \Delta \vec{p} = \Delta(m\vec{v})$ | Linear impulse changes momentum |
| $\vec{\tau}\Delta t = \Delta \vec{L} = \Delta(I\vec{\omega})$ | Rotational impulse changes angular momentum |

## TABLE C.3 ELECTRICITY AND MAGNETISM EQUATIONS

| | |
|---|---|
| $F_E = \dfrac{Kq_1q_2}{r^2}$ | Electric force between two charges |
| $\vec{E} = \dfrac{\vec{F}}{q}$ | Electric field is electric force per charge |
| $P = IV$ | Power is current times the potential difference |
| $I = \dfrac{Q}{\Delta t}$ | Current is charge in motion |
| $\Delta V = IR$ | Ohmic devices provide a linear relationship between voltage difference and current flow |
| $C = \dfrac{Q}{V}$ | Capacitance is the amount of charge per volt |
| $F = q\vec{v} \times \vec{B} = qvB\sin\theta = I\ell B\sin\theta$ | Magnetic force on charge(s) in motion inside of an externally generated magnetic field |
| $B = \dfrac{\mu_0 I}{2\pi r}$ | Magnetic field from a straight line of current |
| $V_{induced} = \text{EMF} = \dfrac{\Delta\Phi_B}{\Delta t}$ | Induced voltage from a change in magnetic flux |

## TABLE C.4 OSCILLATORY AND WAVE EQUATIONS

| | |
|---|---|
| $\omega = 2\pi f$ $\\$ $T = \dfrac{1}{f}$ | Angular frequency (rad/s) Frequency (cycles/s) Period (seconds/cycle) |
| $x(t) = A\cos(\omega t + \Phi)$ | Simple harmonic motion |
| $\omega = \sqrt{\dfrac{g}{L}}$ | Angular frequency for pendula |
| $\omega = \sqrt{\dfrac{k}{m}}$ | Angular frequency for springs |
| $v = f\lambda$ | Wave speed is frequency (set by source) times wavelength in that medium |
| $f_R = \dfrac{(v \pm v_R)f_S}{v \pm v_S}$ | Doppler shift based on relative motion between source and receiver |

## TABLE C.5 ENERGY EQUATIONS

| | |
|---|---|
| $\frac{1}{2}mv^2$ | Kinetic energy |
| $W = \vec{d} \cdot \vec{F} = |\vec{d}||\vec{F}|\cos\theta$ | Work is a force applied over a distance |
| $W_{net} = \Delta K$ | Total work done by all forces changes kinetic energy |
| $W_{nc} = \Delta K + \Delta U$ | Work done by nonconservative forces changes mechanical energy (kinetic plus potential) |
| $U_g = mgh$ | Simplified gravitational potential energy (near Earth's surface) |
| $U_g = -\frac{Gm_1m_2}{r}$ | Universal gravitational potential energy |
| $U_E = qV$ | Electric potential energy is charge times voltage |
| $U_s = \frac{1}{2}kx^2$ | Elastic potential energy (where $x$ is the compression or extension from equilibrium) |

## TABLE C.6 OPTICS EQUATIONS

| | |
|---|---|
| $n = \frac{c}{v}$ | Index of refraction |
| $n_1\sin\theta_1 = n_2\sin\theta_2$ | Snell's law of refraction |
| $\frac{1}{d_i} + \frac{1}{d_o} = \frac{1}{f}$ | Image distance, object distance, and focal length relationship |
| $d\sin\theta = m\lambda$ | Location ($\theta$) of bright fringes (numbered $m$) from two-slit (separation $d$) interference |

## TABLE C.7 THERMODYNAMICS

| | |
|---|---|
| $°K = °C + 273.15$ <br> $°C = \dfrac{5}{9}(°F - 32)$ | Three temperature scales (Kelvin, Celsius, Fahrenheit) |
| $Q = mC\Delta T$ | Heat-temperature rule |
| $\dfrac{1}{2}mv_{avg}{}^2 = \dfrac{3}{2}kT$ | Kinetic theory |
| $\dfrac{3}{2}NkT$ | Internal energy of ideal monoatomic gas |
| $PV = NkT$ | Equation of state for an ideal gas (relating pressure, volume, number of particles, and temperature) |
| $\Delta U = Q - W$ | First law of thermodynamics (internal energy of gas, heat transfer into gas, work done by gas) |
| $\Delta S = \dfrac{Q}{T}$ | Entropy increase when heat $Q$ is added to system |

## TABLE C.8 FLUIDS

| | |
|---|---|
| $F_{buoyant} = V_s \rho_f g$ | Buoyant force in a fluid from the weight of displaced fluid |
| $P + \rho g h + \dfrac{1}{2}\rho v^2$ | Conserved quantity for fluid's pressure, height, and velocity |

## TABLE C.9  MODERN PHYSICS EQUATIONS

| | |
|---|---|
| $\sqrt{\dfrac{1}{1-\dfrac{v^2}{c^2}}}$ | Gamma factor; relativistic length contraction and time dilation due to relative motion |
| $E = mc^2$ | The energy in a system is the mass in the system multiplied by the square of the speed of light ($c$ = speed of light) |
| $KE = hf - W$ | Photoelectric effect; kinetic energy of ejected-by-photon (frequency $f$) electron from metal (work energy $W$) |
| $p = \dfrac{h}{\lambda}$ <br> $E = hf$ | Wave-particle duality <br> Momentum and wavelength <br> Energy and frequency <br> Related by Planck's constant |
| $\Delta x \Delta p \geq \dfrac{h}{4\pi}$ | Heisenberg uncertainty |
| $N(t) = N_0 e^{-t/\tau}$ <br> $t_{\frac{1}{2}} = \tau \ln(2)$ | Quantity, half-life, and mean lifetime for radioactive decay |

# Constants

Table C.10 lists the values of constants commonly used in physics problems.

### TABLE C.10  CONSTANTS

| | |
|---|---|
| $G$, Universal Gravitational Constant | $6.67 \times 10^{-11}$ N • m$^2$/kg$^2$ |
| $g$, local gravitational field | 9.8 (N/kg or m/s$^2$) at Earth's surface |
| $k$, Coulomb's constant $\left(= \dfrac{1}{4\pi\varepsilon_0}\right)$ | $8.89 \times 10^9$ N • m$^2$/C$^2$ |
| $\mu_0$, permeability of free space | $1.26 \times 10^{-6}$ m • kg • s$^{-2}$ • A$^{-2}$ |
| $k$ or $k_B$, Boltzmann's constant | $1.38 \times 10^{-23}$ m$^2$kg/s$^2$K |
| $h$, Planck's constant | $6.63 \times 10^{-34}$ m$^2$kg/s |
| $c$, speed of light (for all massless objects in a vacuum) | $3.00 \times 10^{-8}$ m/s |

# ANSWERS TO CHAPTER REVIEW EXERCISES

## Chapter 2

1. The *net* force determines the winner, not just the single force of tension. Indeed, if tension was the only force (no friction), the forces exerted would be equal. If both teams had the same mass, every game of tug-of-war would be a tie.

2. In all cases, the sensation experienced is in the opposite direction from the acceleration. The person is noting the difference between what her inertia is doing (continuing in a straight line at constant speed) and what the net force is doing to the person.

3. You would explain to James that the forces are indeed the same. However, since Earth is so massive, the resulting acceleration would be much less:

$$a = F/m$$

4. First use Newton's second law to sum the forces in the vertical direction:

$$F_N - F_g = ma_y$$

The sum of the vertical forces is zero because the adult's acceleration is zero:

$$F_N - F_g = 0$$
$$F_N = F_g$$

Since the adult's weight is given, we can plug that into our force equation:

$$F_{g,\text{Earth}} = 800 \text{ N}$$

Since the force due to gravity on Earth is equal to the mass times the strength of gravity on Earth, which is 9.8 N/kg, we can substitute that into the force equation:

$$m(9.8 \text{ N/kg}) = 800 \text{ N}$$

Solve for the person's mass:

$$m = 81.6 \text{ kg}$$

Since mass does not depend on gravity, the adult's mass on the moon is also 81.6 kg. However, his weight will be different on the moon. Next, we know the force due to gravity on the moon is equal to the mass times the strength of gravity on the moon:

$$F_{g,\text{moon}} = ma_{\text{moon}}$$

We are given that gravity is 6 times weaker on the moon than on Earth:

$$a_{\text{moon}} = \frac{g}{6}$$

Plug this into the force equation for the moon:

$$F_{g,\text{moon}} = m\left(\frac{g}{6}\right)$$

Since we know $g$ and calculated $m$ previously, we can solve for $F_{g,\text{ moon}}$:

$$F_{g,\text{moon}} = (81.6 \text{ kg})\left(\frac{9.8 \text{ N/kg}}{6}\right) = 133 \text{ N}$$

Since we found above that $F_N = F_g$ in this situation:

$$F_{N,\text{moon}} = 133 \text{ N}$$

5.  a.  First use Newton's second law to sum the forces in the vertical direction:

$$F_N - F_g = ma_y$$

The sum of the forces in the vertical direction is zero because the book's acceleration is zero:

$$F_N - F_g = 0$$

$$F_N = F_g$$

Since the force due to gravity is equal to the mass times the strength of gravity, we can substitute that into the force equation:

$$F_N = mg$$

Solve for $F_N$:

$$F_N = (2 \text{ kg})(9.8 \text{ N/kg}) = 19.6 \text{ N}$$

b.  First draw a free-body diagram:

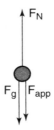

Now use Newton's second law to sum the forces in the vertical direction, where $F_{app}$ is the applied force. The sign on $F_g$ and $F_{app}$ should be the same since they're both applied in the same direction:

$$F_N - F_g - F_{app} = ma_y$$

The sum of the forces in the vertical direction is zero because the book's acceleration is zero:

$$F_N - F_g - F_{app} = 0$$

$$F_N = F_g + F_{app}$$

Since the force due to gravity is equal to the book's mass times the strength of gravity, we can substitute that into the force equation:

$$F_N = mg + F_{app}$$

Solve for $F_N$:

$$F_N = (2 \text{ kg})(9.8 \text{ N/kg}) + 12 \text{ N} = 31.6 \text{ N}$$

c. First draw a free-body diagram:

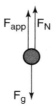

Now use Newton's second law to sum the forces in the vertical direction, where $F_{app}$ is the applied force. The sign on $F_N$ and $F_{app}$ should be the same since they're both applied in the same direction:

$$F_N + F_{app} - F_g = ma_y$$

The sum of the forces in the vertical direction is zero because the book's acceleration is zero:

$$F_N + F_{app} - F_g = 0$$
$$F_N = F_g - F_{app}$$

Since the force due to gravity is equal to the book's mass times the strength of gravity, we can substitute that into the force equation:

$$F_N = mg - F_{app}$$

Solve for $F_N$:

$$F_N = (2 \text{ kg})(9.8 \text{ N/kg}) - 8 \text{ N} = 11.6 \text{ N}$$

d. First use Newton's second law to sum the forces in the vertical direction:

$$F_N - F_g = ma_y$$
$$F_N = ma_y + F_g$$

Since the force due to gravity is equal to the book's mass times the strength of gravity, we can substitute that into the force equation:

$$F_N = ma_y + mg$$

Solve for $F_N$:

$$F_N = (2 \text{ kg})(2 \text{ m/s}^2) + (2 \text{ kg})(9.8 \text{ N/kg}) = 23.6 \text{ N}$$

e. First use Newton's second law to sum the forces in the vertical direction:

$$F_N - F_g = ma_y$$

The sum of the forces is zero because the book's acceleration is zero:

$$F_N - F_g = 0$$

$$F_N = F_g$$

Since the force due to gravity is equal to the mass times the acceleration due to gravity, we can substitute that into the force equation:

$$F_N = mg$$

Solve for $F_N$:

$$F_N = (2 \text{ kg})(9.8 \text{ N/kg}) = 19.6 \text{ N}$$

6. First find the components of the tension force vector given:

$$F_{T,x} = F_T \cos(60°) \qquad F_{T,y} = F_T \sin(60°)$$

Use Newton's second law to sum the forces in the horizontal direction:

$$F_{T,x} - F_f = ma_x$$

Since the frictional force is equal to the normal force multiplied by the coefficient of kinetic friction and since the tension in the horizontal direction was found above, we can substitute these values into the horizontal force equation:

$$F_T \cos(60°) - F_N \mu_k = ma_x$$

Use Newton's second law to sum the forces in the vertical direction:

$$F_N + F_{T,y} - F_g = ma_y$$

The sum of the forces in the vertical direction is zero because the book's vertical acceleration is zero:

$$F_N + F_{T,y} - F_g = 0$$

Since the force due to gravity is equal to the mass times the acceleration due to gravity and since the tension in the vertical direction was found above, we can substitute these values into the vertical force equation:

$$F_N + F_T \sin(60°) - mg = 0$$

$$F_N = -F_T \sin(60°) + mg = (-250)(\sin(60°)) + 85(9.8) = 616 \text{ N}$$

Now we can return to the bolded expression above to determine the acceleration:

$$F_T \cos(60°) - F_N \mu_k = ma_x$$

$$(250)(\cos(60°)) - (616)(0.12) = 85a$$

$$a = 0.60 \text{ m/s}^2$$

# Chapter 3

1.  a.  First draw a line from the position at 0 seconds to the position at 5 seconds.

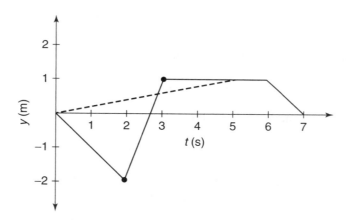

Then calculate the slope of the drawn line to find the average velocity:

$$\frac{(1 \text{ m} - 0 \text{ m})}{(5 \text{ s} - 0 \text{ s})} = 0.2 \text{ m/s}$$

b. First draw a line from the position at 0 seconds to the position at 2 seconds.

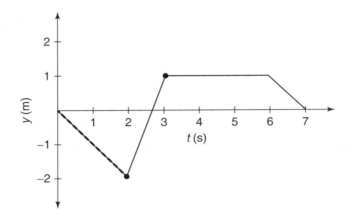

Then calculate the slope of the drawn line to find the average velocity:

$$\frac{(-2 \text{ m} - 0 \text{ m})}{(2 \text{ s} - 0 \text{ s})} = -1 \text{ m/s}$$

c. First draw a line from the position at 2 seconds to the position at 7 seconds:

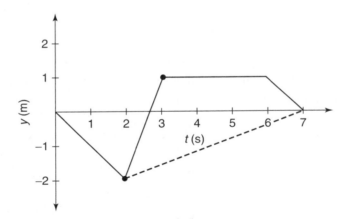

Then calculate the slope of the drawn line to find the average velocity:

$$\frac{(0 \text{ m} - -2 \text{ m})}{(7 \text{ s} - 2 \text{ s})} = 0.4 \text{ m/s}$$

2.  a.  Since the interval from 0 seconds to 2 seconds has a constant slope, the instantaneous slope at the first second will be the same as the slope from 0 seconds to 2 seconds. Calculate the slope from 0 seconds to 2 seconds to find the instantaneous velocity:

$$\frac{(-2 \text{ m} - 0 \text{ m})}{(2 \text{ s} - 0 \text{ s})} = -1 \text{ m/s}$$

b.  Since the interval from 2 seconds to 3 seconds has a constant slope, the instantaneous slope 2.5 seconds into the trip will be the same as the slope from 2 seconds to 3 seconds. Calculate the slope from 2 seconds to 3 seconds to find the instantaneous velocity:

$$\frac{(1 \text{ m} - -2 \text{ m})}{(3 \text{ s} - 2 \text{ s})} = 3 \text{ m/s}$$

c.  Since the interval from 3 seconds to 6 seconds has a constant slope, the instantaneous slope at the fourth second will be the same as the slope from 3 seconds to 6 seconds. Calculate the slope from 3 seconds to 6 seconds to find the instantaneous velocity:

$$\frac{(1 \text{ m} - 1 \text{ m})}{(6 \text{ s} - 3 \text{ s})} = 0 \text{ m/s}$$

3.  Your average velocity is your displacement divided by the total time. Start by calculating the displacement traveled north, which is speed multiplied by time. Be careful to convert minutes into hours!

$$\text{displacement north} = \left( \frac{35 \text{ miles}}{\text{hour}} \right)\left( \frac{1}{3} \text{ hour} \right) = 11.7 \text{ miles}$$

Now calculate the displacement south:

$$\text{displacement south} = \left( \frac{40 \text{ miles}}{\text{hour}} \right)\left( \frac{1}{6} \text{ hour} \right) = 6.7 \text{ miles}$$

Since the displacements were in opposite directions, subtract them:

Net displacement = 11.7 miles north – 6.7 miles south = 5 miles north

Now calculate your average velocity for the entire trip:

$$\text{velocity} = \frac{\text{displacement}}{\text{time}}$$

$$= \frac{5 \text{ miles}}{\left(\frac{1}{3} + 1 + \frac{1}{6}\right) \text{ hours}} = 3.33 \text{ mph northward}$$

4.  Calculate the speed by dividing distance by time. Remember that speed is a scalar quantity because it has no direction.

$$\text{speed} = \frac{\text{distance}}{\text{time}}$$

$$= \frac{(11.7 + 6.7) \text{ miles}}{\left(\frac{1}{3} + 1 + \frac{1}{6}\right) \text{ hours}} = 12 \text{ mph}$$

5.  a.  Since projectile motion problems are symmetrical, the velocity of the rock when you catch it is –20 m/s.

    b.  The velocity of the rock at the highest point is 0 m/s. All turnaround points have an instantaneous velocity of zero in the direction of the turnaround.

    c.  Use the following kinematics equation to find the time in:

$$v_f = v_i + at$$

We know these values:

$$v_f = -20 \text{ m/s}$$

$$v_i = 20 \text{ m/s}$$

$$a = -9.8 \text{ m/s}^2$$

Plug these known values into the kinematics equation and solve for $t$:

$$-20 \text{ m/s} = 20 \text{ m/s} + -9.8 \text{ m/s}^2(t)$$

$$t = 4.1 \text{ seconds}$$

6.  a.  First draw a line from the velocity at 0 seconds to the velocity at 5 seconds. Then calculate the slope of the line you drew to find the average acceleration:

$$\frac{(3 \text{ m/s} - 0 \text{ m/s})}{(5 \text{ s} - 0 \text{ s})} = 0.6 \text{ m/s}^2$$

b.  First draw a line from the velocity at 0 seconds to the velocity at 2 seconds. Then calculate the slope of the line you drew (since the graph is straight here, the lines are the same) to find the average acceleration:

$$\frac{(3 \text{ m/s} - 0 \text{ m/s})}{(2 \text{ s} - 0 \text{ s})} = 1.5 \text{ m/s}^2$$

c.  First draw a line from the velocity at 2 seconds to the velocity at 8 seconds. Then calculate the slope of the drawn line to find the average acceleration:

$$\frac{(0 \text{ m/s} - 3 \text{ m/s})}{(8 \text{ s} - 2 \text{ s})} = -0.5 \text{ m/s}^2$$

7.  a.  Since the interval from 0 seconds to 2 seconds has a constant slope, the instantaneous slope at the first second is the same as the slope from 0 seconds to 2 seconds. Calculate the slope from 0 seconds to 2 seconds to find the instantaneous acceleration:

$$\frac{(3 \text{ m/s} - 0 \text{ m/s})}{(2 \text{ s} - 0 \text{ s})} = 1.5 \text{ m/s}^2$$

b.  Since the interval from 2 seconds to 5 seconds has a constant slope, the instantaneous slope 2.5 seconds into the trip is the same as the slope from 2 seconds to 5 seconds. Calculate the slope from 2 seconds to 5 seconds to find the instantaneous acceleration:

$$\frac{(3 \text{ m/s} - 3 \text{ m/s})}{(5 \text{ s} - 2 \text{ s})} = 0 \text{ m/s}^2$$

c.  Since the interval from 5 seconds to 8 seconds has a constant slope, the instantaneous slope at the seventh second is the same as the slope from 5 seconds to 8 seconds. Calculate the slope from 5 seconds to 8 seconds to find the instantaneous acceleration:

$$\frac{(0 \text{ m/s} - 3 \text{ m/s})}{(8 \text{ s} - 5 \text{ s})} = -1 \text{ m/s}^2$$

8.  The area under the curve equals the net displacement. Multiply the units of each axis to find the units for the area (net displacement in this case):

$$(m/s)(s) = m$$

The area under the curve can be broken up into 2 triangles and a rectangle. Find the area of each shape, and then add them to get the net displacement:

Triangle 1: $(0.5)(3 \text{ m/s})(2 \text{ s}) = 3$ m

Rectangle: $(3 \text{ m/s})(4 \text{ s}) = 12$ m

Triangle 2: $(0.5)(3 \text{ m/s})(3 \text{ s}) = 4.5$ m

$$3 \text{ m} + 12 \text{ m} + 4.5 \text{ m} = 19.5 \text{ m}$$

9.  The change in velocity divided by the change in time is equal to the average acceleration:

$$\frac{(v_f - v_i)}{(t_f - t_i)} = \text{average acceleration}$$

Plug in the given values, and solve for the average acceleration:

$$\frac{(9.0 \text{ m/s} - 25 \text{ m/s})}{(4 \text{ s} - 0 \text{ s})} = -4 \text{ m/s}^2$$

10. Assuming the car is undergoing uniform acceleration, we can use the following equation of motion:

$$\text{displacement} = v_i t + \frac{1}{2} a t^2$$

$$= 25(4) - \frac{1}{2}(4)(16) = 68 \text{ m}$$

# Chapter 4

1.  a.

    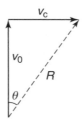

    b.  The resulting speed will be greater because the hypotenuse of a triangle is always greater than its legs.

    c.

    $$\text{Magnitude} = \sqrt{2.6^2 + 1.8^2} = 3.2 \text{ m/s}$$

    $$\theta = \tan^{-1}\left(\frac{1.8}{2.6}\right) = 35°$$

2.  a.

    b.  The resulting speed will be smaller since your rowing speed of 2.6 m/s is now the hypotenuse of the triangle and the resultant is one of the legs.

    c.

    $$\theta = \sin^{-1}\left(\frac{1.8}{2.6}\right) = 44°$$

3.  a.  First calculate the vertical component of the initial velocity:

    $$v_{yi} = v\sin\theta$$

    Plug in the given values to solve for $v_{yi}$:

    $$(15.0 \text{ m/s})\sin(70.0°) = 14.1 \text{ m/s}$$

You know the final velocity is the opposite of the initial velocity. Use the kinematics equation in the vertical direction to solve for time:

$$v_f = v_i + at$$

Plug in the calculated or known values, and solve for $t$:

$$-14.1 \text{ m/s} = 14.1 \text{ m/s} + (-9.8 \text{ m/s}^2)t$$

$$t = 2.88 \text{ s}$$

b.  First calculate the horizontal component of the initial velocity:

$$v_{xi} = v\cos\theta$$

Plug in the given values to solve for $v_{xi}$:

$$(15.0 \text{ m/s})\cos(70.0°) = 5.13 \text{ m/s}$$

You know that the horizontal velocity is constant in flight. So multiply the initial horizontal velocity by the time in flight found in part (a) to get the range:

$$(5.13 \text{ m/s})(2.88 \text{ s}) = 14.8 \text{ m}$$

c.  You know that at the cannonball's highest point, its vertical velocity is 0 m/s. Next, you can use the kinematics equation in the vertical direction to solve for time:

$$v_f = v_i + at$$

You know $v_f$ is 0, and $v_{yi}$ was calculated in part (a). Plug in the calculated or known values, and solve for $t$:

$$0 \text{ m/s} = 14.1 \text{ m/s} + (-9.8 \text{ m/s}^2)t$$

$$t = 1.44$$

Note that this is the answer to part (a) divided by 2 since the flight is symmetrical.

d.

| Time | x-Position | y-Position | $v_x$ Component | $v_y$ Component |
|---|---|---|---|---|
| Initial | 0 m | 0 m | 5.13 m/s | 14.1 m/s |
| At cannonball's highest point | $v_x t$<br>5.13(1.44)<br>= 7.4 m | $v_y t - 4.9t^2$<br>14.1(1.44) − 4.9(2.07)<br>= 10.2 m | 5.13 m/s | 0 m/s |
| As cannonball hits the ground | $v_x t$<br>5.13(2.88)<br>= 14.8 m | 0 m | 5.13 m/s | −14.1 m/s |

e. First find the components of the cannonball's velocity as it hits the ground. Since you are ignoring friction, the final horizontal velocity of the cannonball is the same as the initial horizontal velocity:

$$v_{xi} = v_{xf} = 5.13 \text{ m/s}$$

Since you are ignoring friction, the final vertical velocity of the cannonball is the opposite of the initial vertical velocity:

$$v_{yi} = -v_{yf} = -14.1 \text{ m/s}$$

With the two components of the cannonball's velocity, you can use the Pythagorean theorem to solve for the resulting speed:

$$a^2 + b^2 = c^2$$

Plug in the components to solve for speed:

$$(5.13 \text{ m/s})^2 + (-14.1)^2 = v_R^2$$

$$v_R = 15.0 \text{ m/s}$$

This value is the speed but not the angle. To find the angle of the cannonball, you can use trig:

$$\tan\theta = \frac{\text{opposite}}{\text{adjacent}}$$

$$\tan\theta = \frac{v_{yf}}{v_{xf}}$$

$$\tan^{-1}\left(\frac{v_{yf}}{v_{xf}}\right) = \theta$$

Plug in the components and solve for the angle:

$$\tan^{-1}\left(\frac{-14.1 \text{ m/s}}{5.13 \text{ m/s}}\right) = -70.0°$$

Note that since this is a symmetrical projectile motion problem; the projectile falls down and to the left at the same angle that it was launched up and to the right.

4. Use Newton's version of Kepler's third law:

$$\frac{T^2}{r^3} = \frac{4\pi^2}{GM}$$

You can solve for $M$, which is the mass of the central object, the sun in this case:

$$T = 365 \text{ days} \times \left(\frac{24 \text{ hours}}{1 \text{ day}}\right) \times \left(\frac{3,600 \text{ seconds}}{1 \text{ hour}}\right) = 3.15 \times 10^7 \text{ seconds}$$

$$M = \frac{4\pi^2 r^3}{GT^2}$$

$$M = \frac{4\pi^2 \left(1.50 \times 10^{11} \text{ m}\right)^3}{\left(6.67 \times 10^{-11} \text{ m}^3/\text{kgs}^2\right)\left(3.15 \times 10^7 \text{ s}\right)^2} = 2.01 \times 10^{30} \text{ kg}$$

5. a. Speed is constant, but acceleration is the change in velocity. Velocity includes changes in direction, and circular motion involves constantly changing direction.

b. First, tension is the only force acting in the centripetal direction:

$$F_c = F_T = \frac{mv^2}{r}$$

Next, plug in the values given and solve for velocity:

$$150 \text{ N} = \frac{(0.50 \text{ kg})v^2}{0.25 \text{ m}}$$
$$v = 8.7 \text{ m/s}$$

c. In a horizontally swung circle, the tension force is the only force acting in the centripetal direction, and it is constant. Gravity is not acting in the centripetal direction. However, in a vertically swung circle, gravity and tension are both forces in the centripetal direction. For example, at the top of the circle, tension and gravity are acting toward the center of the circle. Whereas at the bottom of the circle, tension is acting toward the center of the circle, but gravity is acting outward from the circle.

d. The string is most likely to break at the bottom of the circle because the tension force has to counteract gravity in addition to providing the centripetal acceleration. The string would not be most likely to break at the top of the circle because tension benefits from gravity acting in the same direction as tension. To determine the maximum speed, you know that tension and gravity are the only centripetal forces and that tension is positive (inward) and gravity is negative (outward):

$$F_c = F_T - F_g = \frac{mv^2}{r}$$

Since the force due to gravity is equal to the mass times the acceleration due to gravity, you can substitute that into the centripetal force equation:

$$F_T - mg = \frac{mv^2}{r}$$

Next, plug in the values given and solve for velocity:

$$150 \text{ N} - (0.50 \text{ kg})(9.8 \text{ N/kg}) = \frac{(0.50 \text{ kg})v^2}{0.25 \text{ m}}$$

$$v = 8.5 \text{ m/s}$$

# Chapter 5

1.  First, you could push the door directly into the door frame. There would be no resulting torque because the angle between the force and the radius is $180°$. The sine of $180°$ is 0, so there would be no torque. Second, you could pull the door directly out of the door frame. There would be no resulting torque because the angle between the force and the radius is $0°$. The sine of $0°$ is 0, so there would be no torque. A third option would be to push directly on the door hinges. Since the radius is 0 at that location, the torque is 0 as well.

2.  As you can see, the upward force cancels out the downward force, resulting in a net force of zero. Therefore, the center of mass experiences no acceleration. However, both torques are counterclockwise. Therefore, they add together rather than canceling out, causing an angular acceleration to be experienced.

3.  The torque is responsible for changing the diver's angular momentum from the initial zero to what is required during the fall. Assume the flips take place while the diver is at her lowest moment of inertia:

$$L = I\omega$$

$$L = 6 \text{ kg} \bullet \text{m}^2 \times \left( \frac{4.5 \text{ rotations}}{1.5 \text{ seconds}} \right) \times \left( \frac{2\pi \text{ radians}}{\text{rotation}} \right) = 113 \frac{\text{kg} \bullet \text{m}^2}{\text{s}}$$

Note that it would be reasonable to use an average value of $9 \text{ kg} \bullet \text{m}^2$ here as well.

Using this estimate of final angular momentum, you can calculate torque as the rate of change of angular momentum:

$$\tau = \frac{(L - 0)}{t} = \frac{113}{0.15} = 750 \text{ N} \bullet \text{m}$$

4.  a.  First, each force is tangent. So the angle between radius and force is 90 degrees. Note that the three $F$ forces are all exerting counterclockwise (CCW) torques, so they are positive. In contrast, the $2F$ force is exerting a clockwise (CW) torque, so it is negative:

$$\tau_{net} = F(2R) + F(3R) + F(3R) - 2F(3R) = 2FR$$

The positive result indicates the net torque is CCW.

b.  Since you can assume the moment of inertia of the system is 16.0 kg • m²:

$$\tau_{net} = I\alpha$$

$$2FR = 16\alpha$$

$$\alpha = \frac{(2)(24.5)(1.5)}{16} = 4.6 \text{ rad/s}^2$$

c.  Its final angular velocity, $\omega_f$, after 8.0 seconds can be calculated:

$$\omega_f = \omega_i + \alpha t = 0 + (4.6)(8) = 37 \text{ rad/sec}$$

d.  Calculate the total angular displacement of the system:

$$\theta = \omega_i t + \frac{1}{2}\alpha t^2$$

$$\theta = 0 + \frac{1}{2}(4.6)(8)^2 = 147 \text{ radians}$$

5.  a.  Since the rod does not have any angular acceleration, the net torque equals zero:

$$\tau_{net} = I\alpha = 0$$

You must add three torques: the rod, the line, and the block. The torque of the rod's weight and the block's weight are both clockwise (CW), while the line's torque is counterclockwise (CCW):

$$\tau_{rod} + \tau_{block} - \tau_{line} = 0$$

Torque equals the cross product of the radius and the force:

$$\tau = rF\sin\theta$$

Plug this into the net torque equation for each torque:

$$rF\sin\theta_{rod} + rF\sin\theta_{block} - rF\sin\theta_{line} = 0$$

The force the rod is exerting is its weight, and the radius is the center of the rod because it is uniform. Gravity acts perpendicularly to the radius. Plug these values into the net torque equation:

$$\left(\frac{r}{2}\right)mg\sin\theta_{rod} + rF\sin\theta_{block} - rF\sin\theta_{line} = 0$$

The force the block exerts on the rod is the block's weight, and the radius is the whole rod. Gravity acts perpendicularly to the rod. Plug these values into the net torque equation:

$$\left(\frac{r}{2}\right)m_{rod}g\sin90° + rm_{block}g\sin90° - rF\sin\theta_{line} = 0$$

The line exerts a tension force on the rod, and the radius is the whole rod. Tension acts 30° from the rod. Plug these values into the net torque equation:

$$\left(\frac{r}{2}\right)m_{rod}g\sin90° + rm_{block}g\sin90° - rF_T\sin30° = 0$$

Plug in the given values and solve for $F_T$:

$$\left(\frac{0.60\text{ m}}{2}\right)(2.0\text{ kg})(9.8\text{ N/kg}) + (0.60\text{ m})(0.50\text{ kg})(9.8\text{ N/kg}) - (0.60\text{ m})F_T\sin30° = 0$$

$$F_T = 29.4\text{ N}$$

b. By using Newton's second law, you can sum the forces in the horizontal direction acting on the rod. The hinge and the line exert a force in the horizontal direction. The system is not accelerating, so the sum of the forces equals zero:

$$F_{net,x} = ma_x$$

$$F_{T,x} + F_{hinge,x} = 0$$

$$F_{T,x} = F_{hinge,x}$$

Now solve for the horizontal tension force in the line. You found the total tension force in part (a). Solve for $F_{T,x}$:

$$\cos\theta = \frac{F_{T,x}}{F_T}$$

$$F_{T,x} = 20.6\cos30° = 17.8\text{ N}$$

Since the $F_{T,x}$ cancels the $F_{hinge,x}$ as shown above:

$$F_{hinge,x} = 17.8 \text{ N}$$

Solve for the force of the hinge in the vertical direction:

$$F_{T,y} + F_{hinge,y} - Mg - mg = 0$$

$$F_{hinge,y} = (M_{rod} + m_{block})g - F_{T,y} = (2 + 0.5)(9.8) - 20.6\sin30° = 14.2 \text{ N}$$

# Chapter 6

1. a. Momentum is conserved in a system that includes both cars.

   b. A change in momentum due to impulse would have to be calculated in a system that includes only one of the cars.

2. a. A system that includes the object, the spring, and the attached end of the spring has conserved mechanical energy.

   b. In a system that consists of the sliding object only, the work done by the spring would be changing the kinetic energy of the object.

3. Since the mass of the wall is unknown, including the wall in the system would not be appropriate as its changes in energy would not be known. The system should be the cart alone and should treat the forces exchanged with the wall as external.

4. a. The system here is one of the moon alone with Earth's gravity as an external force.

   b. Potential energy is used only if both objects sharing the potential energy are in the system. In this case, the system is the Earth and the moon.

5. a. Neither the sun nor Mercury is an isolated system for this purpose, so the angular momentum is conserved only if both are included in the system.

   b. By treating the torque supplied by the sun's gravity as an external force, the system is just Mercury.

6.

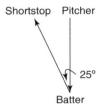

Shortstop   Pitcher

25°

Batter

a. $v_{ix} = 0$          $v_{fx} = -10\sin25° = -4.23$ m/s

$v_{iy} = -8$ m/s          $v_{fy} = +10\cos25° = 9.06$ m/s

b. $F_x t = p_{fx} - p_{ix} = m(-4.23 - 0) = -(0.145)(-4.23) = 0.613$ kg m/s

$F_y t = p_{fy} - p_{iy} = m(9.06 - (-8)) = -(0.145)(17.06) = -2.47$ kg m/s

c. $F_x = \dfrac{0.613}{0.2} = 3.06$ N

$F_y = \dfrac{-2.47}{0.2} = -12.4$ N

Magnitude $= \sqrt{3.06^2 + (-12.4)^2} = 12.8$ N

7.  Since mechanical energy is not conserved through the impact (lost to frictional and other nonconservative forces), you cannot simply conserve mechanical energy through the entire process. Momentum, however, must still be conserved:

$$p_i = p_f$$

$$m_{\text{ball}}v_{\text{ball}} + 0 = (m_{\text{ball}} + m_{\text{clay}})v_{\text{final}}$$

Solving for $v_{\text{final}}$:

$$v_{\text{final}} = \frac{m_{\text{ball}}v_{\text{ball}}}{(m_{\text{ball}} + v_{\text{ball}})} = \frac{0.15v_{\text{ball}}}{(0.15 + 5.6)} = 0.0261v_{\text{ball}}$$

This final, after-impact velocity provides the initial kinetic energy of the upward swing during which mechanical energy is conserved:

$$E_{\text{bottom}} = E_{\text{top}}$$

$$\frac{1}{2}m_{\text{total}}v_{\text{final}}^2 = m_{\text{total}}gh$$

Solve for $h$:

$$h = \frac{v_{final}^2}{2g}$$

Substitute the expression found above for $v_{final}$ in terms of $v_{ball}$:

$$h = \frac{(0.0261 v_{ball})^2}{2g}$$

Plug in $h = 0.15$ m and $g = 9.8$ N/kg, and solve for the initial speed of the ball:

$$v_{ball} = 65.7 \text{ m/s}$$

# Chapter 7

1.

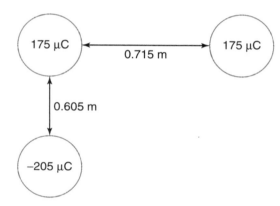

First determine the direction of the individual forces. The upper-left charge will be pushed to the left by the other 175 µC charge. This is a positive charge repelling a positive charge. Call this force $F_1$. The upper-left charge will be pulled downward by the –205 µC charge. This is a negative charge attracting a positive charge. Call this force $F_2$.

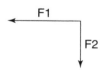

Next, calculate the magnitudes of these forces using Coulomb's law. Note that you are concerned only with magnitude, so no negative signs are used in these calculations:

$$F_1 = \frac{kq_1q_2}{r^2} = \frac{(9 \times 10^9)(175 \times 10^{-6})(175 \times 10^{-6})}{(0.715)^2} = 539 \text{ N}$$

$$F_1 = \frac{kq_1q_2}{r^2} = \frac{(9 \times 10^9)(175 \times 10^{-6})(205 \times 10^{-6})}{(0.605)^2} = 882 \text{ N}$$

Finally, do a vector sum of all the individual forces to find the net effect:

$$F_{net,x} = F_{1,x} + F_{2,x} = (-539 \text{ N}) + 0 = -539 \text{ N}$$

$$F_{net,y} = F_{1,y} + F_{2,y} = 0 + (-882 \text{ N}) = -882 \text{ N}$$

Since the problem requests magnitude and angle:

$$|F_{net}| = \sqrt{F_{net,x}^2 + F_{net,y}^2} = 1{,}030 \text{ N}$$

$$\theta = \tan^{-1}\left(\frac{F_{net,y}}{F_{net,x}}\right) = 58.6°$$

Since both components are negative, the vector is in quadrant III: 58.6° below the negative x-axis or 238.6°.

2.

$$E = qV = (1.6 \times 10^{-19} \text{ C})(9 \text{ V}) - 1.44 \times 10^{-18} \text{ J}$$

or

$$(e)(9V) = 9 \text{ eV}$$

This is the energy lost by the battery per electron (and gained by the circuit).

3. As the sock is tumbling around in the dryer, it may become charged by friction. The charged sock then induces charge separation in your neutral shirt. Since the opposite charges are closer, electrostatic attraction will be stronger than the repulsion between like charges and static cling will be the result.

4. By creating a low-resistance path to the ground, a metal (lightning) rod extending above the highest point of your house provides a safe route for the charges in a lightning strike between the ground and the clouds.

5.

$$\text{Energy} = qV$$

If no charges ($q$) are moving, the voltage difference itself does not indicate any energy being used. Only as charge is transferred across a voltage difference is any energy actually being transferred.

# Chapter 8

1. Item (A): When switch 1 (S1) is open, no current flows from the battery. So neither R1 or R2 gets power.

   Item (B): When switch 2 (S2) is open, no current flows through R1.

   Item (C): When switch 3 (S3) is closed, it creates a short for R1. All the current bypasses R1 in favor of the zero-resistance pathway through switch S3.

| S1 | S2 | S3 | R1 | R2 |
|---|---|---|---|---|
| Open | Open | Open | NO (A) | NO (A) |
| Open | Closed | Closed | NO (A) | NO (A) |
| Closed | Open | Closed | NO (B) | YES |
| Closed | Closed | Closed | NO (C) | YES |
| Closed | Closed | Open | YES | YES |
| Open | Open | Closed | NO (A) | NO (A) |

2. Item (A): These resistors are in series; therefore, the equivalent resistance is the sum of all resistors along the path:

$$25 + 50 + C = 150$$

$$C = 75$$

Item (B): The current coming out of the battery is determined by Ohm's law using the battery's voltage and the equivalent resistance of the entire circuit:

$$V = IR$$

$$300 = I(150)$$

$$I = 2$$

Item (C): The current is the same for all elements in series (Kirchhoff's junction rule).

Item (D): Voltage change across individual resistors is determined by Ohms' law for each resistor: $V = IR$. Results can be double-checked since all voltage changes across resistors in series must add up to the total for the path (Kirchhoff's loop rule).

| Resistor | Resistance ($R$) | Current ($I$) | Voltage Drop ($V$) |
|---|---|---|---|
| $A$ | 25 $\Omega$ | 2 A (Item C) | 50 V (Item D) |
| $B$ | 50 $\Omega$ | 2 A (Item C) | 100 V (Item D) |
| $C$ | 75 $\Omega$ (Item A) | 2 A (Item C) | 150 V (Item D) |
| Battery | N/A | 2 A (Item B) | 300 V |
| Entire circuit equivalent | 150 $\Omega$ | N/A | N/A |

3. Item (A): These resistors are in parallel to each other and to the battery itself. The voltage drop must be the same for each since they are on their own path with the battery (Kirchhoff's loop rule).

Item (B): The current through individual resistors is determined by Ohm's law for each resistor: $V = IR$.

Item (C): For resistor $B$, you can use Kirchhoff's junction rule to determine the current:

$$1.2 = 0.5 + 0.33 + B$$

$$B = 0.37 \text{ A}$$

Item (D): The current through individual resistors is determined by Ohm's law for each resistor: $V = IR$.

Item (E): You can use two ways to find the equivalent resistance. First, you can use $V = IR$ for the current coming out of the battery:

$$9 = 1.2R$$

$$R = \frac{9}{1.2} = 7.5\ \Omega$$

Alternatively, you can use the following:

$$\frac{1}{R_{eq}} = \frac{1}{R_A} + \frac{1}{R_B} + \frac{1}{R_C}$$

$$\frac{1}{R_{eq}} = \frac{1}{18} + \frac{1}{24.3} + \frac{1}{27}$$

$$R_{eq} = 7.5\ \Omega$$

$$R = 7.5\ \Omega$$

| Resistor | Resistance ($R$) | Current ($I$) | Voltage Drop ($V$) |
|---|---|---|---|
| $A$ | 18 $\Omega$ | 0.5 A (Item B) | 9 V (Item A) |
| $B$ | 24.3 $\Omega$ (Item D) | 0.37 A (Item C) | 9 V (Item A) |
| $C$ | 27 $\Omega$ | 0.33 A (Item B) | 9 V (Item A) |
| Battery | N/A | 1.2 A | 9.0 V |
| Entire circuit equivalent | 7.5 $\Omega$ (Item E) | N/A | N/A |

4.

$$RC = \text{ohms} \cdot \text{farad}$$

$$= \left(\frac{\text{volt}}{\text{amp}}\right) \times \left(\frac{\text{coulomb}}{\text{volt}}\right)$$

$$= \frac{\text{coulomb}}{\text{amp}} = \left(\frac{\text{coulomb}}{\frac{\text{coulomb}}{\text{second}}}\right)$$

$$= \text{seconds}$$

# Chapter 9

1.  a.  Use the right-hand rule: thumb to the right (direction of positive charge), pointer
        finger into the page (direction of magnetic field), middle finger points to the top
        of the page (direction of force). As the alpha particle curves around in response
        to the force, the direction of your thumb must also change.

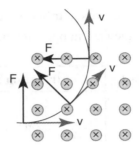

    b.  The magnetic force is acting as a centripetal force:

    $$F_B = F_c$$

    $$qvB = \frac{mv^2}{r}$$

    Solve for $r$:

    $$r = \frac{mv}{qB} = \frac{\left(6.64 \times 10^{-27} \text{ kg}\right)\left(2{,}000 \text{ m/s}\right)}{2\left(1.6 \times 10^{-19} \text{ C}\right)\left(5 \text{ T}\right)}$$

    $$r = 8.3 \times 10^{-6} \text{ m}$$

    c.  Since the charge is opposite in sign, the electron would curve in the opposite
        direction. The radius would be significantly smaller. The force would be
        halved because the charge is halved. However, since the mass of an electron is
        significantly less than that of an alpha particle, the acceleration would be much
        larger, requiring a much smaller radius in order to maintain the same speed as
        the electron curves.

2.  a.  First you must determine the direction of the magnetic fields sourced by both
        currents. By using the right-hand rule, $I_1$ creates magnetic fields coming out
        of the page above it (point $A$) and into the page below it (points $C$, $D$, and
        $E$). Since $I_2$ is pointed in the opposite direction, points $A$, $B$, and $C$ have their

magnetic fields going into the page and point $E$ has its magnetic fields coming out of the page.

Then you determine the magnitude of the magnetic field source for each current:

$$B = \frac{\mu_0 I}{2\pi r}$$

Finally, you add together the two contributions if they are in the same direction and take the difference if they are in opposite directions.

| Location | Contribution from $I_1$ | Contribution from $I_2$ | Total |
|---|---|---|---|
| A | $\frac{\mu_0 2}{2\pi(0.5)}$ <br><br> Out | $\frac{\mu_0 5}{2\pi(1.5)}$ <br><br> In | $1.3 \times 10^{-7}$ T <br> Out of page |
| B | None | $\frac{\mu_0 5}{2\pi(1.0)}$ <br><br> In | $1 \times 10^{-6}$ T <br> Into page |
| C | $\frac{\mu_0 2}{2\pi(0.5)}$ <br><br> In | $\frac{\mu_0 5}{2\pi(0.5)}$ <br><br> In | $2.8 \times 10^{-6}$ T <br> Into page |
| D | $\frac{\mu_0 2}{2\pi(1.0)}$ <br><br> In | None | $4 \times 10^{-7}$ T <br> Into page |
| E | $\frac{\mu_0 2}{2\pi(1.5)}$ <br><br> In | $\frac{\mu_0 5}{2\pi(0.5)}$ <br><br> Out | $1.7 \times 10^{-6}$ T <br> Out of page |

b. Since the current and the magnetic field are at right angles:

$$F = IlB\sin\theta$$

or

$$\frac{F}{l} = IB$$

Determine the direction by the right-hand rule for point $B$:

$$\frac{F}{l} = I_1 B_B = (2)\left(1 \times 10^{-6}\right) = 2 \times 10^{-6} \text{ N/m}$$

$I_1$ is to the right (thumb); the field is into page (finger); and the middle finger (force) is *toward the top of the page*.

Determine the direction by the right-hand rule for point $D$:

$$\frac{F}{l} = I_2 B_D = (5)\left(4 \times 10^{-7}\right) = 2 \times 10^{-6} \text{ N/m}$$

$I_2$ is to the left (thumb); the field is into page (finger); and the middle finger (force) is *toward the bottom of the page*.

Note that the forces are equal and opposite, and they should be based on Newton's third law!

3.  a.  The electric field is the strongest at point $A$ since the field lines are closest together there.

    b.  If you placed a proton at point $B$, it would first encounter point $C$ since positive charges follow the field lines.

    c.  If you placed an electron at point $B$, it would first encounter point $A$ since charges go in the opposite direction to the field lines.

    d.  Point $A$ has the highest voltage. Electric field lines point from high voltage to low voltage.

    e.  Point $D$ has the lowest voltage. Electric field lines point from high voltage to low voltage.

    f.  Somewhere to the left of the picture there are positive charges since they source field lines.

g.

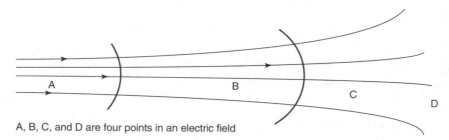

A, B, C, and D are four points in an electric field

h. There are fields in empty space. To have energy or force, though, there must be a charge in the field.

4. Calculate the electric field strength:

$$E = \frac{F}{q} = \frac{\left(3.2 \times 10^{-18}\right)}{\left(1.6 \times 10^{-19}\right)} = 20 \text{ N/C}$$

# Chapter 10

1.

$$\frac{V_{\text{primary}}}{N_{\text{primary}}} = \frac{V_{\text{secondary}}}{N_{\text{secondary}}}$$

$$\frac{7,200}{N_{\text{primary}}} = \frac{240}{N_{\text{secondary}}}$$

$$\frac{N_{\text{primary}}}{N_{\text{secondary}}} = \frac{7,200}{240} = 30$$

Notice from the picture (or look at transformers on your street) that there is a ground wire coming out from the middle of the coil on the secondary side. By connecting the ground wire with one of the other side wires, you obtain 120 volts. By making a circuit with the leftmost and rightmost wires, you get the full 240 volts.

2.   a.   Since the magnetic field and the area are at right angles and since the magnetic field is constant, we can reduce Faraday's law as follows:

$$\text{EMF} = \frac{\Delta \Phi_B}{\Delta t} = \frac{B(\Delta A)}{\Delta t} = \frac{B(l\Delta w)}{\Delta t} = Bl\frac{\Delta w}{\Delta t} = Blv$$

$$\text{EMF} = (0.8)(0.5)(7.5) = 3 \text{ V}$$

In this equation, $A$ is the area in the field, $l$ is the constant length of the bar, and $w$ is the width of the square area receiving the change in flux.

   b.   By Lenz's law, the induced current will be such that it opposes the change in flux. Since the magnetic flux is into the page and is increasing, the induced current will create a magnetic field coming out of the page in the flux area. By the right-hand rule (thumb in the direction of current, fingers showing the direction of fields), the current should move from $b$ to $a$.

   c.   Use Ohm's law to determine the induced current:

$$V = IR$$

$$3 = I(1.5)$$

$$I = 2 \text{ A}$$

Next calculate the magnetic force experienced by this current in the existing magnetic field:

$$F = IlB = (2)(0.5)(0.8) = 0.8 \text{ N}$$

By the right-hand rule (thumb toward top of page, finger into the page, middle finger is to the left), this force is to the left. In order to pull with no acceleration, an equal force to the right is required: 0.8 N to the right.

3.   Multiply wavelength by frequency:

$$(\text{wavelength})(\text{frequency}) = \left(\frac{\text{meters}}{\text{wave}}\right)\left(\frac{\text{waves}}{\text{second}}\right) = \text{m/s}$$

For electromagnetic radiation, the result is always the speed of light: $3.0 \times 10^8$ m/s in a vacuum.

# Chapter 11

1. a. First draw the picture of the standing wave. Each end will be a node. The node-to-node distance is half of a wavelength.

| | String Held Fixed at Both Ends | Frequency | Wavelength |
|---|---|---|---|
| $N = 1$; 1st harmonic; fundamental frequency | | $f = \dfrac{v}{\lambda} = \dfrac{180}{2.7}$ $f = 66.7$ Hz | $\lambda = 2L = 2.7$ m |
| $N = 4$; 4th harmonic; 3rd overtone | | $f = \dfrac{v}{\lambda} = \dfrac{180}{0.675}$ $f = 267$ Hz | $\lambda = \dfrac{1}{2}L = 0.675$ |

   b. Pitch is frequency and is set by the source: 66.7 Hz.

   Wavelength is determined by the wave equation using the speed of sound in air:

   $$\lambda = \frac{v}{f} = \frac{340}{66.7} = 5.1 \text{ m}$$

   c. The loose end will be an antinode, and the fixed end will be a node.

| | String Held Fixed at One End While Loose at the Other | Frequency | Wavelength |
|---|---|---|---|
| $N = 1$; 1st harmonic; fundamental frequency | | $f = \dfrac{v}{\lambda} = \dfrac{180}{5.4}$ $f = 33.3$ Hz | $\lambda = 4L = 5.4$ m |

2. a. $v_R$ is zero since the receiver is not moving. The frequency will be shifted higher since the ambulance is approaching, so use the subtraction operation in the denominator:

   $$f_R = \frac{\left(v \pm v_R\right)f_S}{v \pm v_S} \qquad f_R = \frac{v(780)}{v - 27.5} = \frac{(340)(780)}{340 - 27.5} = 849 \text{ Hz}$$

b. Since the ambulance is now moving away, the frequency will be shifted lower. So the addition operation in the denominator is the one to use:

$$f_R = \frac{v(780)}{v + 27.5} = \frac{(340)(780)}{340 + 27.5} = 722 \text{ Hz}$$

3.  a.
$$dB = 10\log\left(\frac{I}{I_0}\right) = 10\log\left(\frac{10^{-7}}{10^{-12}}\right) = 50 \text{ dB}$$

b.
$$dB = 10\log\frac{I}{I_0}$$

$$90 = 10\log\left(\frac{x}{10^{-12}}\right)$$

$$10^9 = \frac{x}{10^{-12}}$$

$$x = 10^{-3} \text{ W/m}^2$$

4.  a. Half the distance from peak to trough: 5 cm

b. Time for one complete cycle: 4 s

c. $\dfrac{1}{T} = 0.25$ Hz

d. $2\pi f = \dfrac{\pi}{2}$ rad/s

e. $\omega = \sqrt{\dfrac{k}{m}}$

$$\frac{\pi}{2} = \sqrt{\frac{k}{m}}$$

$$\left(\frac{\pi}{2}\right)^2 m = k$$

$k = 18.5$ N/m

f. $\dfrac{1}{2}kx^2$ maximizes when $x = A$. (There is no kinetic energy when $x = A$.)

$$E = \frac{1}{2}kA^2 = \frac{1}{2}(18.5)(0.05)^2 = 0.023 \text{ J}$$

g. Points of maximum kinetic energy occur whenever the position goes through zero. These locations have the steepest slopes (highest velocities).

h. Points of maximum elastic potential energy occur whenever the displacement is maximum or minimum (peaks and troughs on the graph).

i. $F = kx$ maximizes when $x = A$.

$F = kA = 18.5(0.05) = 0.925 \text{ N}$

j. When velocity is at its maximum, $x = 0$ and all energy will be kinetic. Using the total energy from part (f):

$$E = \frac{1}{2}mv^2 + \frac{1}{2}kx^2 = \frac{1}{2}mv_{max}^2 = \frac{1}{2}kA^2 = E_{tot} = 0.023 \text{ J}$$

Solve for $v_{max}$:

$$v_{tot} = \sqrt{\frac{2E_{tot}}{m}} = \sqrt{\frac{2(0.023)}{7.5}} = 0.078 \text{ m/s}$$

k. Since the sinusoid starts at a maximum, you can use the sine function (or a cosine function with an appropriate phase shift):

$$y(t) = A\sin(\omega t) = 0.05\sin\left(\frac{\pi t}{2}\right)$$

5. 1.5 picometers $= \lambda = 1.5 \times 10^{-12}$ m

This is so short that it falls into the gamma ray range of electromagnetic radiation:

$$f = \frac{v}{\lambda} = \frac{c}{\lambda} = \frac{\left(3.0 \times 10^8\right)}{\left(1.5 \times 10^{-12}\right)} = 2 \times 10^{20} \text{ Hz}$$

# Chapter 12

1. First draw a ray that passes from the focal point on the same side of the convex lens as the object. It will refract to be horizontal. Then draw a ray that is horizontal; it will refract to pass through the focal point on the opposite side. Note that these two rays do not cross, so they must be traced back to find a point where they appear to come from (a virtual image).

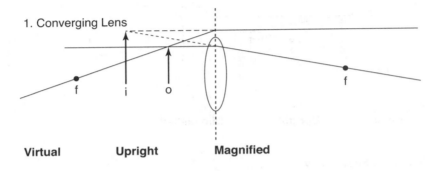

First draw a ray directed toward the center of curvature ($R$ in this picture) of the convex mirror. It will reflect directly back since this ray is normal to the surface. Then draw a horizontal ray; it will reflect directly away from the focal point (half of the radius). Note that the two rays do not cross, so they must be traced back to find a point from where they appear to come from (a virtual image).

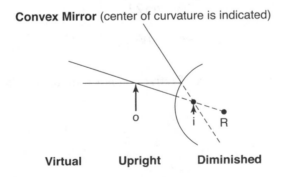

First draw a horizontal line, it will refract as if it came from the focus on the same side as the object of the diverging lens. Then draw a line straight through the center of the lens; it will not refract at all. Note that the two rays do not cross, so they must be traced back to find a point from where they appear to come from (a virtual image).

1. Diverging Lens

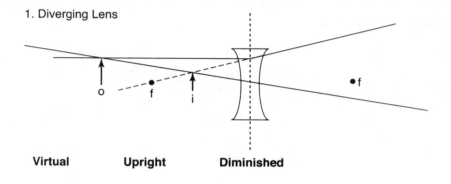

**Virtual**        **Upright**        **Diminished**

2.  a.  Start with Snell's law:

$$n_1 \sin\theta_1 = n_2 \sin\theta_2$$

The index of refraction for air is approximately 1. Plug in the given values:

$$1\sin40° = n_2\sin22°$$

$$n_2 = \frac{\sin40°}{\sin22°} = 1.7$$

$$N_2 = \sin(40)/\sin(22) = 1.7$$

There are no units as index of refraction is a ratio of speeds.

b.

$$n = \frac{c}{v}$$

$$v = \frac{c}{n} = \frac{3.0 \times 10^8}{1.7} = 1.8 \times 10^8 \text{ m/s}$$

c. To find the critical angle, set the angle to 90° in the faster medium because critical angles exist only on the slower side:

$$n_1 \sin\theta_1 = n_2 \sin\theta_2$$

$$1.7\sin\theta_c = 1.47\sin90°$$

$$\sin\theta_c = \frac{1.47}{1.7}$$

$$\theta_c = 60°$$

d. Total internal reflection occurs. There is no solution for a refracted ray in this situation, so all the energy must be reflected.

3. a. Since $\lambda f = v$ and the frequency does not change upon entry into a new medium:

$$\lambda_n = \frac{\lambda}{n}$$

$$\lambda_n = \frac{650 \text{ nm}}{1.6} = 406 \text{ nm}$$

b. Remember the phrase "low to high: phase shift pi." Since the oil is a slower medium, the reflected light is phase shifted by pi.

c. Remember the phrase "high to low: phase shift no." Since the water is a faster medium, the reflected light is not phase shifted.

d. Two rays are interfering. The ray bouncing off the air-oil interface does not travel through the oil. The ray bouncing off the oil-water interface travels through the oil layer twice: once on the way down and once on the way up.

e. You want the wave to be out of phase for destructive interference. Since the two reflection events as described above already have the two waves out of phase, you need the path length difference to be exactly one additional wavelength so they meet out of phase again:

$$\text{Thickness} = \frac{\text{Path length difference}}{2} = \frac{406 \text{ nm}}{2} = 203 \text{ nm}$$

f. The very edges of the oil layer will be dark since the two waves are out of phase from the reflections alone.

g. Since phase shift upon reflection depends on only the relative speeds of the media, the answer to f. does not depend on the wavelength.

4. a. Bright spots are located at point $x$:

$$x = \frac{m\lambda L}{d}$$

Solve for $d$, which is the distance between the slits:

$$d = \frac{m\lambda L}{x}$$

$$d = \frac{(1)(650 \times 10^{-9})(5)}{0.01} = 3.25 \times 10^{-4} \text{ m}$$

b. Green has a shorter wavelength than red. Smaller $\lambda$ produces smaller $x$. Therefore, the bright spots would be closer together.

c. Bigger $d$ produces smaller $x$. So the bright spots would move closer together.

d. Bigger $L$ produces bigger $x$, but the angle of interference maximum remains the same. So the bright spots would move farther apart.

5. Sound waves are longitudinal. As such, they do not have a separate direction for amplitude. (Their amplitudes are in the direction of travel.) So sound waves do not have a direction to polarize.

6. Polarized sunglasses block light that is polarized horizontally. This is especially advantageous in situations where a lot of surface glare is coming off the ground into a person's eyes (near snow, water, etc.). Reflections are preferentially polarized parallel to the surface. Although polarized sunglasses will simply dim the unpolarized light in a similar fashion to regular sunglasses, polarized sunglasses will block most or all of the glare.

# Chapter 13

1. First the water must be raised by 10 degrees Celsius to the boiling point:

$$Q = mc\Delta T$$

$$Q = (500 \text{ gram})(1 \text{ cal/gram}°)(10°) = 5{,}000 \text{ calories}$$

This leaves 2,500 calories out of the 7,500 to change the phase of the water into gas:

$$Q = mL$$

$$2{,}500 = x(540 \text{ cal/g})$$

$$x = \frac{2{,}500}{540} = 4.6 \text{ g}$$

So 4.6 grams of water will be turned to steam.

2. Start with the ideal gas law:

$$PV = NkT$$

Using the conversions of 1 atmosphere = 105 pascals; 1 liter = 0.001 m³; and K = °C + 273:

$$\left(2.5 \times 10^5\right)(3 \times 0.001) = N\left(1.38 \times 10^{-23}\right)(85 + 273)$$

$$N = 1.5 \times 10^{23} \text{ molecules}$$

3. Start with the relationship between temperature and energy:

$$\text{Average kinetic energy} = \frac{1}{2}mv_{\text{avg}}^2 = \frac{3}{2}kT$$

Room temperature is around 20 degrees C, or 293 K:

$$\text{Average kinetic energy} = \frac{3}{2}\left(1.38 \times 10^{-23} \text{ J/K}\right)(293 \text{ K})$$

$$= 6.06 \times 10^{-21} \text{ J}$$

Calculate the mass of nitrogen gas ($N_2$). Since this is a diatomic gas, 1 mole of nitrogen gas has 28 grams of nitrogen (twice the average mass number). This is the mass for an Avogadro's number of molecules. One molecule would have a mass of:

$$\frac{28 \text{ grams}}{6.02 \times 10^{23}} = 4.64 \times 10^{-26} \text{ kg}$$

You previously calculated the average kinetic energy:

$$\frac{1}{2}mv^2 = 6.06 \times 10^{-21} \text{ J}$$

Solve for $v$:

$$v = \sqrt{\frac{(2)(4.04 \times 10^{-21})}{4.64 \times 10^{-26}}} = 511 \text{ m/s}$$

4.

Heat lost by hot object(s) = Heat gained by cool object(s)

$$m_{hot}C_{hot}\Delta T_{lost} = m_{cold}C_{cold}\Delta T_{gained}$$

Let $x$ be the final equilibrium temperature:

$$(550)(4.19)(97 - x) = (750)(0.902)(x - 22)$$

Note that degrees Celsius can be used here since you are using the *change* in temperature, and the change in Celsius is the same as the change in Kelvin.

Solve for $x$:

$$223,500 - 2,304.5x = 676.5x - 14,883$$

$$x = 80°$$

5. First the ice must absorb enough heat to raise it to the melting point (9 degrees):

$$Q = mC\Delta T$$

$$Q = (150)(2.1)(9)$$

$$Q = 2,835 \text{ J}$$

Next use the heat of fusion to melt the ice:

$$Q = mL$$

$$Q = (150)(335)$$

$$Q = 50{,}250 \text{ J}$$

Finally use the specific heat of water to heat the water to 21 degrees:

$$Q = mC\Delta T$$

$$Q = (150)(4.18)(21)$$

$$Q = 13{,}167 \text{ J}$$

Add up all the heat to find the total:

$$2{,}835 \text{ J} + 50{,}250 \text{ J} + 13{,}167 \text{ J} = 66{,}252 \text{ J}$$

6.  Since the number of molecules in the balloon is not changing, you know the following from the ideal gas law:

$$\frac{P_i V_i}{T_i} = \frac{P_f V_f}{T_f}$$

Plug in the given quantities. Remember to include initial atmospheric pressure and to convert temperature to the Kelvin scale. Note that you do not need to convert the volume to $m^3$ since you are comparing ratios:

$$\frac{\left(10^5\right)(220)}{(28 + 273)} = \frac{(35{,}000)x}{(-11 + 273)}$$

$$\frac{\left(2.2 \times 10^7\right)}{301} = \frac{\left(3.5 \times 10^4\right)x}{262}$$

$$x = 547 \text{ liters}$$

7.

$$\text{Ideal efficiency} = 1 - \frac{T_L}{T_H} = 1 - \frac{11 + 273}{175 + 273} = 0.366 = 37\%$$

In actuality, some heat will be generated internally. The real efficiency will be lower as less work will be done since some of the energy taken in is converted to this internal heat.

8. Even if the energy absorbed during the day by Earth was completely emitted at night (i.e., no global warming), the net entropy of the universe is increasing as there are more infrared photons emitted than there are higher-frequency photons absorbed. Sunlight peaks in the visible range, but Earth is emitting much lower, infrared radiation. Since infrared radiation has less energy per photon, there must be more of them emitted. The net effect is to take a more organized type of energy (few photons) and spread it out in a more unorganized way (more photons). Life itself is not actually using energy. It is taking in a more organized type of energy and discarding an equal amount of energy, but that discarded energy is in a greater variety of states (more disorganized). So it is more accurate to say that life is an entropy-increasing process rather than an energy-consuming one!

# Chapter 14

1. Pressure is due to the weight of a column of fluid above a given surface area:

$$P = \rho g h = (1.3)(9.8)(20) = 255 \text{ pascals}$$

2. Assume there is laminar flow:

$$A_1 v_1 = A_2 v_2$$

If the cross-sectional area is halved, the flow rate must double. Now use the fact that $v_2 = 2v_1$ in the Bernoulli equation:

$$P_1 + \rho g h_1 + \frac{1}{2}\rho v_1{}^2 = P_2 + \rho g h_2 + \frac{1}{2}\rho v_2{}^2$$

Since the height is not changing and you know the relationship between the velocities:

$$P_1 + \frac{1}{2}\rho v_1^2 = P_2 + \frac{1}{2}\rho(v_1)^2$$

$$P_1 - P_2 = \frac{3}{2}\rho(v_1)^2$$

The pressure drops by $\frac{3}{2}\rho(v_1)^2$, where $v_1$ is the initial flow rate.

3.

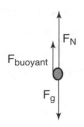

$$F_{net} = ma$$

$$F_N + F_{buoy} - mg = 0$$

$$F_N = -F_{buoy} + mg$$

Use Archimedes rule for buoyant force:

$$F_{buoy} = V_s \rho_f g$$

First calculate the volume in m$^3$:

Volume $= (215 \text{ mm} \times 102.5 \text{ mm} \times 65 \text{ mm})(\text{m}/(10^3 \text{ mm}))^3 = 1.4 \times 10^{-3} \text{ m}^3$

Next, we can use this volume and the density of water ($1,000 \text{ kg/m}^3$) to determine the buoyant force:

$$F_{buoy} = (0.0014)(1,000)(9.8) = 13.7 \text{ N}$$

Substituting into the expression for normal force above:

$$F_N = -F_{buoy} + mg = -13.7 + 5(9.8) = 35.3 \text{ N}$$

4. As the train pulls air along with it, the train increases the flow rate of the air near it. Since the Bernoulli equation remains constant, this increase in flow rate lowers the pressure:

$$P_1 + \rho g h_1 + \frac{1}{2}\rho v_1{}^2$$

Objects nearby are pushed by the higher pressure of the air farther away from the train toward the lower-pressure area where the air is moving quickly.

5. A solid piece of metal does, of course, sink. This is because the weight of the water displaced by the metal is much less than the weight of the metal itself, so the buoyant force is not enough to counteract gravity. A ship, however, is mostly air by volume. The ship displaces so much water that only a portion of it need be under the waterline, actually displacing water, in order to displace enough water to have a buoyant force equal to the weight of the ship itself.

6. As the velocity of the air over the top of the umbrella increases, it lowers the air pressure there, according to the Bernoulli Principle, where the following expression is conserved:

$$P_1 + \rho g h_1 + \frac{1}{2}\rho v_1{}^2$$

In this equation, $h$ is basically the same. Since $v$ increases, $P$ must decrease.

Since the pressure below the umbrella is now higher, there is net upward force. This is the same principle that is partly responsible for the lift force on an airplane's wings.

7. For something to float, the net force must equal zero:

$$F_{net} = ma$$

$$F_{buoy} - mg = 0$$

$$F_{buoy} = mg$$

Rewrite this relationship in terms of volumes and densities:

$$F_{buoy} = V_{submerged}\, \rho_{water}\, g = mg = V_{total}\, \rho_{ice}\, g$$

Rearrange to get the ratio of volumes:

$$\frac{V_{submerged}}{V_{total}} = \frac{\rho_{ice}}{\rho_{water}} = \frac{900}{1,000} = 0.9$$

So 90% of the ice is submerged.

# Chapter 15

1.  First find the difference in energy levels:

$$E_3 - E_1 = \left(\frac{-13.6}{3^2}\right) - \left(\frac{-13.6}{1^2}\right) = -1.51 + 13.6 = 12.1 \text{ eV}$$

Next use Planck's formula for the energy of a photon:

$$E = hf$$

$$f = \frac{E}{h}$$

$$f = \frac{12.1 \text{ eV} \times \left(1.6 \times 10^{-19} \text{ J/eV}\right)}{6.63 \times 10^{-34} \text{ J} \cdot \text{s}} = 2.92 \times 10^{15} \text{ Hz}$$

2.  The conjugate variable for angular momentum is angular position. According to the Heisenberg uncertainty principle, as the precision becomes more accurate for angular momentum, the angular position about the nucleus becomes more uncertain. This is one way of thinking about the electron orbitals: since the angular momentum is well defined, the angular position must be undefined.

3.  a.  There is no way to know. The half-life is a statistical statement that can be applied to large numbers of atoms. However, the time of a single atom undergoing radioactive decay is not predictable, just like knowing the average lifespan of human beings cannot predict when a particular person will die.

    b.  Since the mass of a sample is proportional to the number of atoms, use the formula for radioactive decay:

$$N(t) = N_0 e^{-t/\tau}$$

$$.75 = 12 e^{-t/\tau}$$

Where $\tau = t_{1/2}/\ln(2) = 5730/\ln(2) = 8267$ years

$$.75 = 12e^{-t/8267}$$

$$.0625 = e^{-t/8267}$$

$$\ln(.0625) = -t/8267$$

$$-2.77 = -t/8267$$

$$t = 33,900 \text{ years}$$

c.

$$N(t) = N_0 e^{-t/\tau}$$

$$N(t) = 10^9 e^{-\frac{5000}{10863}} = 10^9 e^{-.46}$$

$$= 6.3 \times 10^8 \text{ atoms}$$

4. Use the Compton formula for momentum:

$$p = \frac{h}{\lambda}$$

Also use the average wavelength of a visible photon, which is $\lambda = 565$ nm:

$$p = \frac{6.63 \times 10^{-34}}{565 \times 10^{-9}} = 1.17 \times 10^{-27} \text{ kg} \bullet \text{m/s}$$

That's a very hard-to-notice momentum!

5. Use deBroglie's version of the Compton wavelength:

$$\lambda = \frac{h}{p}$$

Calculate the momentum of the car:

$$p = mv$$

$$p = (1,400 \text{ kg}) \times \left(\frac{55 \text{ miles}}{\text{hour}}\right) \times \left(\frac{1,609 \text{ meters}}{\text{mile}}\right) \times \left(\frac{1 \text{ hour}}{3,600 \text{ seconds}}\right)$$

$$p = 34,000 \text{ kg} \bullet \text{m/s}$$

Determine the wavelength:

$$\lambda = \frac{6.63 \times 10^{-34}}{34{,}400} = 1.93 \times 10^{-38} \text{ m}$$

That's an extremely-hard-to-notice wavelength!

6. Charge is quantized in units of $1.6 \times 10^{-19}$ C. This means all freestanding occurrences of charge must be an integer number of this quantum unit. Therefore, an isolated charge of $1.1 \times 10^{-19}$ C cannot exist in isolation anywhere in the universe.

7. A beta particle (electron) is emitted when one of the neutrons in the nucleus decays into a proton, raising the atomic number:

$$^{127}_{55}\text{Cs} \rightarrow {}^{127}_{56}\text{Ba}^+ + e^-$$

# Appendix A

1. Since the up and down vectors are equal in magnitude but opposite in direction, they cancel out each other. Looking in the horizontal direction, let's say right is positive and left is negative. The resultant vector is –1 units:

$$5 - 5 = 0 \text{ units}$$

$$6 - 7 = -1 \text{ units}$$

2. Using trigonometry, you must find the supplementary angle of 145 degrees for the triangle:

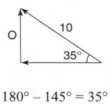

$$180° - 145° = 35°$$

Then using trig values and the given values, you can find the components.

*y*-component:

$$\sin\theta = \frac{\text{opposite}}{\text{hypotenuse}}$$

$$\text{opposite} = 10\sin(35°) = 5.74 \text{ units}$$

*x*-component:

$$\cos\theta = \frac{\text{adjacent}}{\text{hypotenuse}}$$

$$\text{adjacent} = -10\cos(35°) = -8.19 \text{ units}$$

Note that the *x*-component is negative since it is to the left.

3. First by using Pythagorean's theorem, you can calculate the magnitude of the resultant (*H*):

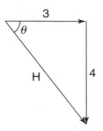

$$3^2 + 4^2 = H^2$$

$$H = 5 \text{ units}$$

To find the direction of the vector, you can use trigonometry and the given values:

$$\tan\theta = \frac{\text{opposite}}{\text{adjacent}}$$

$$\tan\theta = \frac{-4}{3}$$

$$\theta = \tan^{-1}\left(\frac{-4}{3}\right) = -53.1°$$

This means the direction of the vector is −53.1° from the positive $x$-axis.

4.  a.

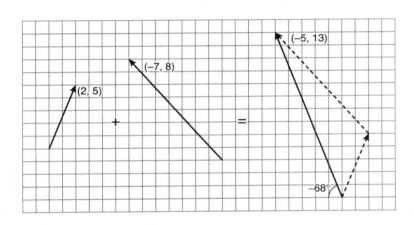

b.

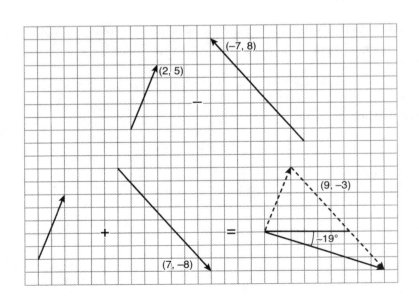

c. For part (a) to get the $x$- and $y$-coordinates of the resultant, you must add the $x$-components and the $y$-components of each vector. For the $x$-component:

$$2 + -7 = -5$$

For the $y$-component:

$$5 + 8 = 13$$

To put the vector into polar notation, you can use Pythagorean's theorem and trigonometry:

$$(-5)^2 + (13)^2 = R^2$$

$$R = 13.9 \text{ units}$$

$$\tan \theta = \frac{y}{x}$$

$$\tan \theta = \frac{13}{-5}$$

$$\theta = \tan^{-1}\left(\frac{13}{-5}\right)$$

$$\theta = 69°$$

The resultant vector is 13.9 units and 69° from the negative $x$-axis, which is 111° (in quadrant II).

For part (b) to get the $x$- and $y$-coordinates of the resultant, you must subtract the $x$-components and $y$-components of each vector. For the $x$-component:

$$2 - (-7) = 9$$

For the $y$-component:

$$5 - 8 = -3$$

To put the vector into polar notation, you can use the Pythagorean theorem and trigonometry:

$$(9)^2 + (-3)^2 = R^2$$

$$R = 9.5 \text{ units}$$

$$\tan \theta = \frac{y}{x}$$

$$\tan \theta = \frac{-3}{9}$$

$$\theta = \tan^{-1}\left(\frac{-3}{9}\right)$$

$$\theta = -18.4°$$

The resultant vector is 9.5 units and −18.4° from the positive $x$-axis (in quadrant IV).

5.   a.   First, to get the $x$- and $y$-coordinates of the resultant, you must add all the $x$-components and $y$-components of each vector. For the $x$-component:

$$A_x + B_x + C_x = 9 + (-4) + 5 = 10$$

For the $y$-component:

$$A_y + B_y + C_y = 2 + (-4) + (-3) = -5$$
$$\vec{A} + \vec{B} + \vec{C} = (10, -5)$$

To put the vector into polar notation, you can use the Pythagorean theorem and trigonometry:

$$(10)^2 + (-5)^2 = R^2$$

$$R = 11.2 \text{ units}$$

$$\tan \theta = \frac{y}{x}$$

$$\tan \theta = \frac{-5}{10}$$

$$\theta = \tan^{-1}\left(\frac{-5}{10}\right)$$

$$\theta = -26.6°$$

Therefore, $\vec{A} + \vec{B} + \vec{C} = 11.2$ units and is $-26.6°$ from the positive $x$-axis (in quadrant IV).

b. First, to get the $x$- and $y$-coordinates of the resultant, you must plug in the values for all the $x$-components and $y$-components of each vector. For the $x$-component:

$$A_x + B_x - 2C_x = 9 + (-4) - 2(5) = -5$$

For the $y$-component:

$$A_y + B_y - 2C_y = 2 + (-4) - 2(-3) = 4$$

$$\vec{A} + \vec{B} - 2\vec{C} = (-5, 4)$$

To put the vector into polar notation, you can use the Pythagorean theorem and trigonometry:

$$(-5)^2 + (4)^2 = R^2$$

$$R = 6.4 \text{ units}$$

$$\tan\theta = \frac{y}{x}$$

$$\tan\theta = \frac{4}{-5}$$

$$\theta = \tan^{-1}\left(\frac{4}{-5}\right)$$

$$\theta = -38.7°$$

Note that the components indicate a quadrant II vector (negative $x$, positive $y$). Therefore, you must locate this angle in the second quadrant by adding $180°$ to the calculated value:

Therefore, $\vec{A} + \vec{B} - 2\vec{C} = 6.4$ units and is $38.7°$ above the negative $x$-axis (or simply $141.3°$).

# INDEX